D0079502

In the Wake of Galileo

In the Wake of Galileo

Michael Segre

Rutgers University Press
New Brunswick, New Jersey

Library of Congress Cataloging-in-Publication Data

Segre, Michael, 1950–
In the wake of Galileo / Michael Segre.
p. cm.
Includes bibliographical references and index.
ISBN 0-8135-1700-1 (cloth) ISBN 0-8135-1701-X (pbk.)
1. Science—Italy—History. 2. Galilei, Galileo, 1564–1642.
3. Italy—Intellectual life—1789–1900. 4. Science—Historiography.
I. Title.
Q127.I8S44 1991
509.45—dc20 90-24573
CIP

British Cataloging-in-Publication information available

Frontispiece: Portrait of Galileo by V. Sustermans, c. 1636 (Deutsches Museum, Munich)

Manufactured in the United States of America

Contents

List of Illustrations

Foreword

I. BERNARD COHEN

Recent years have seen a proliferation of scholarly studies on Galileo and the birth of a "Galileo industry" that is in numbers equal to, if it does not surpass, the "industries" associated with Newton, with Darwin, and with Einstein. In this context we should perhaps say "rebirth" or even "rejuvenation" since Galileo has been the subject of emulation, mythologizing, and study ever since his own lifetime. Known to generations of physicists and philosophers as inventor of the method of experiment in science, Galileo was boldly presented by Alexandre Koyré as a Neoplatonist who eschewed experiment. Hailed as a radical and an innovator in science and method, he has recently been portrayed (by A. Carugo, A. C. Crombie, and William A. Wallace) as one whose method was derived from Jesuit teachers whose notebooks he studied as a youth. Stillman Drake's discovery and interpretation of Galileo manuscripts not used by Antonio Favaro in his monumental edition indicate that the true role of experiment (a topic explored by Drake, Ronald Naylor, and Winifred L. Wisan) in the development of Galileo's thoughts about motion is neither as completely negative as Koyré believed nor as direct and simple as many writers have alleged. R. S. Westfall has challenged the traditional idea of Galileo as "pure" discoverer by showing that a leading source of motivation for new discoveries was not to vindicate the Copernican hypothesis, as has traditionally been assumed, but rather to advance the patronage important to his career. Other studies on Galileo have been concerned with philosophical issues (e.g., those of Maurice A. Finocchiaro, Dudley Shapere, and Michael Segre) and there have been new viewpoints with respect to the hoary question of Galileo and the Church, ranging from Giorgio de Santillana's *Crime of Galileo* to the mild overview of Jerome L. Langford and, most recently, the bold and daring new hypothesis of Pietro Redondi.

In the book before us, Michael Segre has set himself three tasks not yet addressed in full in the current or past welter of Galileo literature. These are to examine the social matrix of Galileo's life and thought; to explore the work of the group of followers or disciples of Galileo; and to study the influence of Galileo in Italy and the state of Italian science during the quarter-century or so after his death. Most histories of science, as Segre notes, omit much of the post-Galilean Italian science when portraying the seventeenth century, confining their attention for the late 1600s to England and France, as if Italian science had come to an end with Galileo. For no other reason, here are grounds for welcoming this new study. On the score of disciples, we have an interesting problem. Viviani, Galileo's major disciple and advocate and his earliest biographer, did not know what to call the master's followers. In a manuscript cited by Segre, he shows his perplexity: Should he call them *scolari* (pupils), *seguaci* (followers), or *discepoli* (disciples)? All three terms are relevant but have different connotations. The subject is one of major interest since we know in other cases how such a network of followers enables a scientist's ideas and influence to spread. Newton, for example, built up such a group, many of whom he helped to place in important positions; their number includes David Gregory, J. T. Desaguliers, Colin Maclaurin, Edmond Halley, and others. This feature of the spread of Newtonian ideas was one of the significant contributions to scholarship in Frank Manuel's *Portrait of Isaac Newton,* a feature of that work often neglected in discussions of the challenging Freudian interpretations of Newton's life.

Segre portrays for us the scientists in Italy who were part of the Galileo movement and who even used the term "school" of Galileo to describe themselves. This aspect of Galilean science owes much to Favaro who, in the National edition of Galileo's works, gave brief notices of individuals associated with Galileo. In 1945, Giorgio Abetti brought attention to this subject in his *Amici e nemici di Galileo.* Paolo Galluzzi and Maurizio Torrini have begun a magnificent edition of *Opere dei discepoli di Galileo Galilei.* But no one until now has tried to assess the significance of this group as a whole, both in relation to the post-Galilean development of science and to the ways in which a scientific movement arises and declines.

Through its clear and balanced picture of the essential features of Galileo's scientific achievements, of his trial before the court of the Inquisition, and of his influence on other scientists, this volume serves as a very good introduction to Galileo. But it does much more than that. Segre shows how Galileo's contemporaries and later admirers contributed to the mythology surrounding this hero of science.

Take, for example, the famous story of dropping weights from the Leaning Tower of Pisa. It is full of interest, not just for what it claims for Galileo, but for the way the details and their significance change in the telling. First of all, the texts themselves record the fact that Galileo did indeed perform experiments (if not public demonstrations) from a tower. In his notes and in the published account, Galileo refers to "a tower," but does not specify which one. Viviani mentions specifically the Leaning Tower, "the campanile of Pisa," a detail that seems to lend verisimilitude to the story. Second, a critical examination of the texts is revealing. Viviani's account states that in the experiment bodies "of the same material" but "of unequal weight . . . all moved at the same speed." This is the version often told in textbooks. But Galileo's own account, published in *Two New Sciences,* reads differently. The weights, he says, followed one another separated by "two finger-breadths." Anyone who has ever dropped weights from the Leaning Tower or made a similar experiment elsewhere will agree that Galileo's account is exact and plausible and that Viviani's is not. Furthermore, this slight difference contains an important clue to the role of air and the fact that the experiment was not carried out in a vacuum.

Such mythmaking about Galileo's life and work (and the consequent lack of interest in his social matrix and his followers) demands attention, and it is surprising that no one has done so before Segre. But in demythologizing Galileo, Segre performs an even greater service to the history of science. He reminds us that good history is the best way to pay tribute to great science: real heroes need no myths.

Preface and Acknowledgments

The Galileo story seems never to lose its fascination, and recent Galilean studies have produced new, interesting, and at times highly significant and controversial contributions. These studies reviewed many aspects of Galileo's life and work, such as his early notes, his working papers on projectile motion, his rhetoric, and the perpetually open issue of his trial. The proliferation of Galilean literature is mainly an outcome of the fact that Galileo became a myth even in his own day and that much energy has been spent over the past three centuries in either strengthening or denouncing this myth. There is a general feeling that the subject has been more or less exhaustively covered.

Curiously, however, recent literature in the history of science has almost totally disregarded one aspect of Galilean studies: Galileo's followers. Although much new work has dealt with the social and institutional background of the Scientific Revolution, for instance, particularly topics such as the rise of the Royal Society of London, little attention has been paid to events during the same period in Italy, as if science ended there after Galileo. Yet post-Galilean science in seventeenth century Italy produced intrinsically important achievements by such scientists as Bonaventura Cavalieri, Evangelista Torricelli, and Giovanni Alfonso Borelli, and was marked by the prominence of Vincenzio Viviani, who had an unmatched international reputation as a scientist and whose efforts were invaluable for the preservation of Galileo's legacy. It also included the activities of the Accademia del Cimento, an institution that had significant formative influence on modern science. A study of Galileo's impact on his contemporaries and successors could thus not only enhance the interpretation of his work but also our understanding of the roots of the Scientific Revolution.

Most of our knowledge of Galileo's followers dates back a century, to the time of Antonio Favaro. Favaro's contribution was remarkable but incomplete, considering the quantity of material available, and in any case ought to be reviewed. A number of recent works do deal with Galileo's followers but are restricted to particular topics, such as Cavalieri's theory of indivisibles or the Accademia del Cimento. To the best of my knowledge no general treatment exists of the interaction between the various members of the "Galilean School," but a useful step in that direction is the beginning of the systematic publication of their correspondence. However, most of the papers of Galileo's followers—comprising hundreds of volumes of manuscripts—are still unpublished, and their accomplishments to a great extent remain to be explored.

In the present study I have tried first of all to paint a general picture of Italian science in the generation after Galileo, relying on both recently published documents and unpublished material. As far as I know, no such presentation has previously appeared in the modern history of science. I also take a novel approach in Galilean studies, seeking in the work of these scientists new insights on Galileo's science and methodology.

Galileo's methodology has been, and still is, a central and controversial issue. In particular, a widespread historiographic tradition depicts Galileo as the first true empiricist, a claim that has perplexed several leading historians and is the subject of continuing debate. The works of Galileo's followers may be an appropriate context for further consideration of the question. One minor but symbolic example of this tradition in popular literature concerns the truth of the story of the Leaning Tower experiment. It is seldom mentioned that Galileo's supposed exploit at the top of the Tower only became known for the first time twelve years after his death in an account by his pupil, Viviani, whose papers were not examined, until my own recent research, for his motives in telling the story. A study of the work of Galileo's followers indicates that the issue of Galileo's empiricism is more questionable than implied by its simplistic presentation in the literature, both popular and scholarly.

More generally, the Galileo myth has given him the peculiar status of official martyr-saint of science. This myth imputes the "collapse" of Italian science during the second half of the seventeenth century solely to the Inquisition. Yet the works and papers of Galileo's followers show that the decline of Italian science was more gradual and complex than is traditionally thought. The Church bears some blame, of course, but intrinsic factors also played a part: Galileo's followers had inherited onerous problems and lacked adequate scientific and mathematical tools to solve them. Sociopolitical factors, too, contributed, and Galileo himself was perhaps partly responsible. He was so busy with his cultural campaign that he neglected to train scientists who could carry on his enterprise after his death. In Tuscany—the stronghold of Galilean science—the fault was also shared by the Medici patrons who applied a cultural policy not always in the best interests of science.

Despite the enormous literature on Galileo and the predominant tendency of modern history of science to concentrate on social aspects, I have not found a single work—not even in Italian—giving an overall context for Galileo's work. I have therefore begun this book with a general sketch of Galileo's time. This rather unorthodox introduction makes no claim to completeness or scholarship. Its sole purpose is to acquaint the reader with some concepts not easy to find in the available literature.

The first chapter of the book also is somewhat introductory, briefly describing Galileo's life and work. Here, I have highlighted certain details necessary for a better understanding of his heritage.

Chapter 2 is historiographic and focuses on Galileo's representation as an empiricist, which started in the time of Galileo's followers, was amplified during the eighteenth and nineteenth centuries, and continued until Galileo's demythologization and remythologization by modern history of science.

The next three chapters deal with the work of and interaction between various members of the "Galilean School" in the period preceding Torricelli's death. Among other things, I cite some unpublished material, in particular Nardi's *Scene*—a rich and interesting treatise of general knowledge still in manuscript form in the National Library of Florence.

Chapters 6 and 7 deal with the period following Torricelli's death in Tuscany, during which science generally languished, although much was done to preserve Galileo's heritage. The dominant figure in this period was Viviani, and chapter 7 deals with what may be regarded as his heroization of Galileo. In my article, "Viviani's Life of Galileo," on which this chapter relies, I claimed that Viviani emphasized the empirical aspects of Galileo's work. Maurice Finocchiaro pointed out to me that a careful reading shows, rather, that Viviani attributed to Galileo a judicious combination of empirical observation and theoretical speculation. It is true that Viviani did also allude to Galileo's commitment to the mathematical and rationalist ideal, but his elegant and elaborate style—like Galileo's—can be interpreted in more than one way, allowing more than one image of Galileo. In my opinion, both the empiricist and judicious Galileos are equally correct readings from Viviani. What is certain is that the empiricist version predominated in later literature and had its origin in Viviani.

The view of science as primarily an empirical enterprise is the subject of chapter 8, which surveys the activities of the Accademia del Cimento. This chapter also deals with the (partly alleged) interference of the Church in the work of Galileo's followers. Some thoughts concerning possible causes of the decline of Tuscan science are presented in the Epilogue.

My book can therefore be described as a general study of science and culture in seventeenth-century Italy. It is not technical and does not claim to contribute to the present knowledge of Galilean physics or astronomy. It endeavors, rather, to tell the story of a community of scientists illuminated by a great scientific figure but not yet ready to embrace the conceptual changes amounting to a scientific revolution. Since most of the material on Galileo's followers is still buried in hundreds of unpublished manuscripts, the story is far from complete. It is more of a mosaic, showing, perhaps, how a myth was created by a scientific movement in decline. I have attempted to combine an "internal" intellectual viewpoint with an "external" socio-institutional one, without getting too involved with epistemology or sociology. I hope simply to show the reader that scientific life continued in Italy in the generation after Galileo and that his followers held conflicting views about both science in general and the Galilean enterprise in particular. I hope also to further the demythologization of Galileo.

I owe much gratitude to many friends and institutions who helped me: my teacher Joseph Agassi encouraged and advised me from the very beginning and contributed much to the improvement of this book. Also, the book would probably not have come into being without the assistance and encouragement of I. Bernard Cohen. I am indebted, too, to Menso Folkerts, for his constant support and his comments, and to Eugenio Garin, Maurice A. Finocchiaro, William Shea, and Ezio Raimondi. Gabriele Bickendorf, Karin Figala, Steven Harris, Antoni Malet, Ivo Schneider, and Pamela Smith read and commented on early drafts, Carlo Maccagni and Richard Lorch were always at my disposal with aid and good advice, and Stella Rasetti enlightened me on some aspects of Cavalieri's theory of indivisibles. Devorah Bar-Zemer and Norman Rudnick patiently edited the book.

The Scuola Normale Superiore of Pisa, the Domus Galileana in Pisa, the reading room of the Biblioteca Nazionale Centrale in Florence, the Istituto e Museo di Storia della Scienza in Florence, the University of Munich, and the Deutsches Museum proved to be invaluable for my research. I am particularly indebted to Heribert M. Nobis for putting at my disposal the Deutsche Copernicus-Forschungsstelle at the Deutsches Museum, where the major part of this book was written, and to Franca Principe who helped me to find and supplied several of the illustrations.

Some parts of the book rely on previously published articles, and I thank the journals that granted me permission to use them. Part of the first chapter appeared in my "Galileo as a Politician" (*Sudhoffs Archiv*). Chapter 2 is a development of "The Role of Experiment in Galileo's Physics" (*Archive for History of Exact Sciences*). Chapter 7 makes extensive use of my "Viviani's Life of Galileo" (*Isis*).

Translations in the text are my own, unless otherwise stated. Although my translations are sometimes a little free, quotations are almost always given in their original language in the endnotes.

Abbreviations

ACG–Antonio Favaro, *Amici e corrispondenti di Galileo.* 3 vols. Reprint. Florence: Salimbeni, 1983.

AIMSSF–*Annali dell'Istituto e Museo di Storia della Scienza di Firenze.*

BB–*Bullettino di Bibliografia e di Storia delle Scienze Matematiche e Fisiche.*

BJHS–*British Journal for the History of Science.*

Carteggio–Paolo Galluzzi and Maurizio Torrini (eds.), *Le opere dei discepoli di Galileo Galilei, Edizione Nazionale: Carteggio.* 2 vols. Florence: Barbèra, 1975–1984.

Crew and de Salvio–Galileo Galilei, *Dialogues Concerning Two New Sciences.* Translated by Henry Crew and Alfonso de Salvio. New York: Macmillan, 1914.

DSB–Charles C. Gillispie (ed.), *Dictionary of Scientific Biography.* 16 vols. New York: Scribners, 1970–1980.

Gal. MSS–The collection of Galilean manuscripts in the Biblioteca Nazionale Centrale, Florence.

OG–Antonio Favaro (ed.), *Le opere di Galileo Galilei, Edizione Nazionale.* 20 vols. Florence: Barbèra, 1890–1909. Reprinted 1929–1939, 1964–1966, 1968.

OT–Gino Loria and Giuseppe Vassura (eds.), *Le opere di Evangelista Torricelli.* 3 vols. Vols. 1–3, Faenza: Montanari, 1919. Vol. 4, Faenza: Lega, 1944.

Saggi–*Saggi di naturali esperienze fatte nell'Accademia del Cimento.* Florence, 1667.

TT–Giovanni Targioni Tozzetti, *Notizie degli aggrandimenti delle scienze fisiche accaduti in Toscana nel corso di anni LX. del secolo XVII.* 3 vols. Florence: 1780. Reprinted. Bologna: Forni, 1967.

Note: I have used Crew and de Salvio's English translation of Galileo's *Two New Sciences,* rather than Stillman Drake's more recent and better translation (see bibliography), because the latter was unfortunately not available to me when I was writing this book.

In the Wake of Galileo

INTRODUCTION: Galileo's Time

Galileo Galilei was born in Pisa in 1564—three days before Michelangelo's death and about two months before Shakespeare's birth—and died in Arcetri, near Florence, in 1642, a little less than a year before Newton's birth.[1] Whereas Michelangelo marks the end of the Italian Renaissance, Shakespeare symbolizes the beginning of the most important period in the renaissance of English culture, and Newton the beginning of the modern scientific and technological era. Galileo lived and worked in a period of transition for Europe in general and for Italy in particular, an interval characterized by a profound moral, religious, and social crisis. The period followed the great geographical discoveries and the Lutheran Reformation; Europe was gradually losing its feeling of centrality and the Catholic Church most of its authority and prestige. It was the divide between two major European intellectual traditions: the Renaissance and modern rationalism.

Surprisingly, the vast literature on Galileo and his intellectual revolution pays relatively little attention to some meaningful aspects of the complex background to his life and work, such as the political, cultural, and social environment in which he lived and which inevitably affected his activities. Most writings about Galileo, for instance, mention that he served the Republic of Venice and the Tuscan Court, that he taught at universities and was active in academies, and that he had controversies with the Church. But very few take account of the often intricate historical processes affecting these institutions and Galileo's involvement with them.

A few pages are inadequate to describe the general context of Galileo's work. However, some basic features of the time may be useful to the reader who is not familiar with the history of post-Renaissance Italy. Experienced historians of science who find this introduction superfluous can skip it and begin the book at the next chapter.

The Council of Trent (1545–1563) was undoubtedly the main historical event on the eve of Galileo's birth. It initiated the Counter-Reformation, the Church of Rome's response to the Lutheran Reformation by closing its ranks through calls for increased discipline and piety. Galileo would become enmeshed in this process when he was accused, reproached, and finally tried and condemned for having broken the rules instituted by the Council.[2]

The Council of Trent and the Counter-Reformation brought in their wake the Thirty Years' War (1618–1648). Italy shared the decline of the Catholic world and of the Spanish Empire, but was largely untouched by the war. It remained divided into small states and did not escape the severe economic crisis that followed the war. The decline was felt in all its aspects in Galileo's homeland, Tuscany and its capital, Florence, once a leading center of the Renaissance.

Florence and Tuscany

Florence had emerged as an important center of commerce, learning, and the arts between the thirteenth and fourteenth centuries, despite frequent disputes between the Guelfs and the Ghibellines—the two opposing factions sympathizing, respectively, with the Pope and the Emperor. In 1406 its forces conquered Pisa, extending its rule over a large part of Tuscany, and in the fifteenth century it reached relative political stability under the rule of the Medici family.

Florence became the center of the Renaissance, the homeland of some of the greatest artists of the period, including Giotto, Masaccio, Donatello, Ghiberti, Brunelleschi, and Alberti. It was in Florence that Dante, Petrarch, and Boccaccio elevated the Tuscan dialect to the status of an all-Italian language, and Machiavelli introduced his conception of the modern state.[3]

The splendor of Florence culminated during the reign of Lorenzo the Magnificent (between 1469 and 1492), the most brilliant Medici to rule Florence and a patron of the arts and letters. His death—a few months before America was discovered—was followed by political unrest and gradual economic and cultural decline. Thus, on the eve of Galileo's birth Tuscany was no longer what it had been during the fourteenth and fifteenth centuries, although it was experiencing relative prosperity, thanks to the authoritarian (and sometimes ruthless) but conscientious administration of Grand Duke Cosimo I de' Medici (1519–1574). Cosimo began his reign in 1537 as Duke of Florence, annexed to his duchy most of the remaining parts of today's Tuscany, including Siena, and in 1569 was declared by the Pope to be Grand Duke of Tuscany.

The reigns of the grand dukes Franceso I, Ferdinand I, and Cosimo II, who followed Cosimo I from 1574 to 1621, were also relatively prosperous. It was in this period that Galileo spent his formative years in Pisa, and it was Cosimo II who, in 1610, invited Galileo to the Tuscan court as his philosopher and mathematician and encouraged his scientific work.

Nevertheless, the death of Cosimo II, in 1621, marked the beginning of the long, final decadence of the Medici and their state, which lasted over a century and directly affected Galileo as the court scientist. The Thirty Years' War was underway, and the Medici who ruled Tuscany after Cosimo II were unable to run the country effectively and to deal with the consequences of the war. When Cosimo II died, his eldest son, Ferdinand II, was only ten years old, and the grand duchy was temporarily ruled by a council of regents headed by the Grand Duchesses Christina, Cosimo's mother, and Maria Maddalena, his widow. The two grand duchesses allowed the country to sink into an economic crisis and submitted to the influence of the clergy, who took advantage of their inability to govern and gradually took over control of the country.

The political situation of Tuscany became even more complicated after 1623, when Cardinal Maffeo Barberini became Pope (Urban VIII). His election was a result of his sympathy toward France, a sympathy that throughout his Papacy was to raise tensions between the Church and Spain and, in consequence, between the Papal States and Tuscany, which at that time was still a vassal princedom of Spain. Urban's main concern was to turn his state into a leading European power, and throughout his tenure he persisted in attempting to establish the hegemony of the Church and the Papal States over Italy.[4] The personal hatred between the Barberini family and other Italian ruling families, including the Medici, contributed to the general political tension. A long period of unrest followed, ending in a war, from 1642 to 1644, between the Papal States and a league of Italian states that included Tuscany. It was during this politically troubled period, from 1624 to 1632, that Galileo wrote his *Dialogue Concerning the Two Chief World Systems,* and its publication led to his trial and condemnation by the Roman Inquisition.

The political tension became particularly strong in 1632, shortly before Galileo's trial, when the army of the Protestant King of Sweden, Gustavus II Adolphus, swept through Europe and reached the south of Germany. He was supported by Cardinal Richelieu, the (Catholic) minister of Louis XIII. Since the Pope sympathized with France, he was publicly accused by the Spanish Ambassador in Rome of protecting heretics. Many cardinals sided with Spain and requested a change in papal policy. Although no direct evidence links the political controversy to Galileo's misfortune, it seems probable that had the Grand Duke of Tuscany been stronger, or relations between Tuscany and the Papal States been more amicable, the Grand Duke might have more easily protected his Court Scientist. And, had the Pope been under less pressure, he might have been able to be more lenient with Galileo.

International politics was only one of the contributors to the weakening of Tuscany. Tuscany also suffered periods of famine and plague, for example, from 1630 to 1633, which spread over most of the Italian peninsula (and was beautifully portrayed two centuries later by Italy's greatest modern writer, Alessandro Manzoni, in *The Betrothed*).[5] Although Tuscany was well organized to fight this onset of plague, Ferdinand II was generally a poor administrator and took more interest in his favorite hobby, science, than in his country. It was only thanks to the help of his talented and hard working younger brother, Prince Leopold de' Medici (1617–1675), that the country did not collapse economically and politically. Leopold also was an amateur scientist, and his patronage was vital to Galileo's followers. Leopold continued, in practice, to govern Tuscany even after Ferdinand's death, during the reign of Ferdinand's son, Cosimo III (1642–1723), who was as ineffectual as his father and held nominal power for no less than fifty-three years. After Leopold's death, in 1675, Tuscany became totally dependent on the Holy Roman Empire. Cosimo's son, Gian Gastone (1671–1737), was the last Medici to rule Tuscany. All these events naturally also affected cultural and scientific developments.

Cultural Institutions

During the sixteenth century Italian culture was the domain of various institutions, such as universities, religious orders, academies, botanical gardens, and courts. These institutions differed widely. The universities and orders, for instance, were both religious and well established; the others were less established and not directly religious but often more liberal and open to innovation. Galileo and his followers were active mainly within the universities, academies, and courts.

Italian universities flourished during the late Middle Ages but ceased to be centers of innovation long before the sixteenth century. The legal Latin term *universitas* described a group of persons with a common aim or function, having the right to sue or to be sued. Universities, whose origin is somewhat obscure, were corporate associations of scholars—teachers and students—later called *studia generalia.* Among the earliest European universities were the medical school of Salerno in southern Italy and the Universities of Bologna (for law) and Paris.[6]

Universities normally had four faculties—arts, medicine, law, and theology. What we now call science was cultivated and taught in the faculties of arts and medicine. Cosmology (the description of the universe), natural philosophy (the study of nature), and many branches of mathematics were taught in the faculty of arts. Yet, by the sixteenth century universities often differed greatly from each other but were in most cases educational institutions with programs of study that remained to a large extent scholastic.

Religious orders also often had schools of different levels, which were sometimes receptive to new ideas such as Galileo's, depending on the ideology of the order. Established after the founding of the Society of Jesus in 1534, Jesuit schools and colleges (of which there were hundreds, all over Europe), in particular, played an important role in the scientific revolution. The leading Jesuit university—the Roman College—was by the end of the sixteenth century one of the most important scientific centers in Europe.

While religious institutions may have been conservative and unreceptive to certain types of knowledge, many scholars and humanists were able to develop their talents under the patronage of wealthy enlightened princes. Many Italian and non-Italian rulers encouraged the arts, letters, and sciences—including occult sciences—and had in their service engineers, artists, physicians, theologians, philosophers, musicians, astrologers, and other such craftsmen or scholars. Leonardo da Vinci, for instance, spent most of his life in various Italian courts and in the service of the French kings. Kepler was Mathematician to the Holy Roman Emperor, and Galileo was Philosopher and Mathematician to the Grand Duke of Tuscany—one of a series of court mathematicians who served at the Tuscan court during the sixteenth and seventeenth centuries.

Sometimes rulers supported entire academies, which, unlike the universities, were private, informal, learned societies created during and after the Renaissance, often as groups of followers of one notable humanist. Their aim was to develop and promote the literature, arts, and sciences that emerged during the Renaissance. Generally they did not offer their members economic support, but acted as confraternities and, like the universities, had formal structures similar to those of the guilds (the term "academy" was also frequently used to describe a *studium*). They had some form of written constitution, held more or less regular meetings, and in most cases were supported by a patron.

It is difficult to draw a general picture of an academy in Galileo's day since, unlike the universities, they were not subject to a common rule.[7] The hundreds of academies active at that time differed widely in their purposes, methods, means, number of members, and duration. Many of those founded during the middle of the sixteenth century were literary academies aiming to protect and encourage the vernacular language. They had unusual and suggestive names, reflecting the literary refinement and sense of humor of their founders, such as *Crusca, Rozzi, Intronati, Umidi,* meaning, approximately, "chaff," "the rude," "the shaken," and "all wet," respectively. Although the members of these academies were expected to express their ideas in some sort of literary—spoken or written—form, the ideas themselves were not limited to literature and could touch upon any field of knowledge from theology to natural philosophy. They had a broad and heterogeneous public and were thus the best audience for Galileo and his followers. Many of these literary academies, whose contribution included the translation of theological treatises into the vernacular, had a short life and did not survive the death of their founders. But they were receptive to new ideas, like Galileo's, and helped to further scientific innovation. There also were strictly "scientific" academies, such as the Lincean (the Academy of the "Lynx-eyed"), founded in 1603 in Rome and active until 1630, and the Tuscan Accademia del Cimento (*cimento* meaning experiment), intimately involved with the Tuscan court between 1657 and 1667. Both these academies strongly supported the work of Galileo or his followers. More detail on the activities of the academies will appear later, but let us concentrate here on the relations between the Medici—in particular the grand dukes, Galileo's patrons—and the various institutions, and with culture in general.

Medici Patronage

From the beginning of their dominance in the fifteenth century, the Medici encouraged many types of cultural activities, which developed Florence into a leading European center of arts, letters, and sciences. Cosimo de' Medici the Elder (1389–1464), a wealthy banker and the first Medici to rule Florence, was an enlightened patron who surrounded himself with humanists, opened libraries, and promoted the arts. He paid agents to look for classical manuscripts in the Arab world, admired Plato, and established the Platonic Academy, an informal group of intellectuals led by Marsilio Ficino (1433–1499), Plato's translator. This was a first step toward the broad cultural policy elaborated by the later Medici.[8]

Cosimo's patronage was continued and extended by his grandson, Lorenzo the Magnificent, who fostered the arts, letters, philosophy, and occult sciences, opened his court to well-known humanists, enriched the Medicean libraries with manuscripts and printed books, and stimulated the translation of classical manuscripts. Lorenzo, who had been a pupil of Ficino, promoted the work of the Platonic Academy and protected Pico della Mirandola (1463–1494) when the church accused the humanist and Cabalist of heresy.[9] In 1472 Lorenzo combined the rival Tuscan universities of Pisa (founded in 1343) and Florence (founded in 1349) in order to concentrate higher education in Pisa. As a result, most of the Florentine studium was dissolved, and Pisa became the leading Tuscan university. But unrest that followed Lorenzo's death adversely affected the University of Pisa, and in the 1520s it ceased its activity. It was later reactivated by Grand Duke Cosimo I.

Although Cosimo I was not an intellectual, he inaugurated a new era of Medici patronage. He was well aware of the importance of cultural patronage as a source of prestige, and even power, and raised culture to the status of a major state project. He let Giorgio Vasari (1511–1574), the famous mannerist painter, architect, and writer who designed the Uffizi Gallery in Florence, supervise the state's artistic interests. In general, Cosimo advanced Tuscan cultural institutions and at the same time subjected them to state control.[10]

In 1543 Cosimo reopened the studium of Pisa. When Siena was annexed to the Tuscan state in 1557, its studium became the second fully active Tuscan university. Cosimo also established chairs in theology and astrology at the studium of Florence, but Pisa retained its primary importance and prestige.

The studium of Pisa in Galileo's time was a typical Italian university. It had three faculties: law, arts, and theology. The faculty of law was by far the largest, with about two-thirds of the university's students; the faculty of arts, in which Galileo studied, included medicine and mathematics; the faculty of theology was very small.

Cosimo intended to turn the Pisan studium into the country's main study center but first had to gain full control. He therefore gave it a new set of statutes, and from then on the state's administration increasingly interfered in the program of studies. The power of the rector—official head of the university—was gradually decreased, and actual power was given to the overseer (*provveditore*), the local director of education, who was subject to the orders of the grand duke.[11] The number of ecclesiastics teaching in Pisa was reduced and their power limited.[12] The studium became a State University in every sense; it trained state officers, the state turned to it for advice, and students wishing to graduate had to take an oath of loyalty to the state.[13] But Cosimo's plans were even more ambitious, and he also invited famous scholars from abroad in the hope of developing Pisa into a leading European university.[14]

Cosimo was probably even more eager to control the academies than the universities, because academy members were intellectuals, free thinkers, and individualists who often regarded him as the usurper of the Florentine Republic and objected to his rule. He therefore began to apply to academies a policy like that imposed on the University of Pisa, encouraging their activities, to enhance his derived prestige, but also restraining their independence. When, in 1540, a group of Florentine merchants established the literary Accademia degli Umidi, Cosimo persuaded its members to abandon it and to join the Accademia Fiorentina (Florentine Academy), which he founded, supported, and controlled. The Accademia degli Umidi was dissolved three months after its start, and the Florentine became the major Tuscan academy.

As Eric Cochrane, a historian of Italian academies, points out, the Florentine Academy was an innovation in the cultural life of Italy.[15] It was not the usual learned society but rather something like an "open university," seeking not only to cultivate the arts and letters but also to propagate them through public lectures. The academy head, or consul, received the same status and salary as the rector of the Florentine studium. The Florentine Academy thus replaced, in practice, the University of Florence, and Cosimo made sure that its officers were dependent upon him.[16] In 1582 several members of the Florentine Academy resigned because they found it too structured and founded the less formal Accademia della Crusca, which soon became the leading Italian language academy.

The Florentine Academy may be regarded as an early instance of modern planned cultural policy. Cosimo had converted the potentially hostile Accademia degli Umidi into a quasi-university, which he could control. The Florentine Academy also helped considerably to reestablish Florence as a major European cultural center and to enable the Medici to exercise direct influence over culture. It is in this "academic" Academy that Galileo began his academic career by lecturing on the *Divine Comedy*.

The Florentine was paralleled by other academies, which were given an official, or semiofficial, status, such as the Accademia del Disegno, a fine-arts academy founded in 1562 (Vasari was one of its founders). The literature does not always distinguish between the academies and the studium.

Cosimo also nurtured new fields, for example, by establishing a chair of botany in Pisa in the early 1540s and creating a botanical garden, probably the earliest in Europe (another botanical garden was created more or less at the same time in Padua). Both the chair and management of the garden were assigned to the botanist Luca Ghini (1490–1556), thereby initiating a tradition of naturalistic studies in Tuscany.

Cosimo thus went beyond the cultural patronage of his predecessors, creating a distinct cultural policy that provided a more solid and regular support to culture. Nevertheless, the policy had the drawback of making culture more and more dependent on the state, a disadvantage that became significant later, particularly in the second half of the seventeenth century when the state's interference was sometimes more harmful than beneficial. Meanwhile, in the decades that followed the death of Cosimo I Tuscan culture flourished again, and it was within the framework of his cultural policy that his grandson, Cosimo II, invited Galileo to the Tuscan Court.

Before presenting Galileo—his life, work, and relations with his patrons and followers—let me describe briefly the state of the art on the eve of his birth in the fields to which he was to contribute, philosophy and mathematics.

Philosophy and Mathematics

Galileo was initially referred to as a mathematician and later insisted on being regarded also as a philosopher. Today we would describe him rather as a physicist and astronomer because the modern meanings of the terms "philosophy," "mathematics," "physics," and "astronomy" are different from those of Galileo's time.

Physics then was part of philosophy, more precisely the part of Aristotelian natural philosophy that studied living and nonliving things. Philosophy was defined in the earliest edition of the *Vocabolario degli Accademici della Crusca* (1612) as "the *true* knowledge of natural things, and of divine and human ones, as far as man is able to grasp" (I emphasize the word "true" because the question of the truth of a hypothesis played a determining role in Galileo's campaign in favor of heliocentrism).[17] Physics was simply the "science of nature" and almost totally devoid of mathematics, and one of Galileo's main efforts was an attempt to unify the two. Despite Galileo's efforts, however, this definition of physics persisted in the 1680 edition of the *Vocabolario,* nearly forty years after his death.

What was the science of nature?

Aristotle's physics is broad and complex and can hardly be presented in a few words. Generally speaking, Aristotle pictured the world as a nest of spheres all centered about the earth and carrying the sun, the moon, and the five planets known at that time. He conjectured that different laws governed the celestial and terrestrial ("sublunary") regions. He adopted Empedocles' view that matter on earth was composed of the four basic elements, water, earth, air, and fire. Each of the elements had a "natural tendency"—water and earth toward the center of the universe, air and fire away from it—and these tendencies determined the physical behavior of a substance. Aristotle said that the speed of fall of a body was proportional to its weight, denied the existence of a vacuum and of atoms, and held that matter is continuous. (This, of course, is an oversimplification of Aristotle's views.)

A popular view presents Aristotelianism as a single philosophy and Aristotelian natural philosophers as relying on references to Aristotle's writings rather than empirical evidence in trying to understand and describe nature. It is true that some of the philosophers rebutted by Galileo and his followers resisted empirical evidence, as in the well-known case of the Paduan philosopher Cremonini who refused to look into Galileo's telescope. But Aristotelianism and its adherents were by no means so single-minded; many different interpretations and various schools of Aristotelianism arose during and after the Renaissance. Some Aristotelian natural philosophers, for instance, made early attempts to understand nature in terms of mathematics, and others emphasized the importance of empirical observation.[18]

Some earlier attempts to apply mathematics to Aristotle's physics had been made in the fourteenth century, chiefly by Nicole Oresme (ca. 1320–1382), a French philosopher and mathematician who served Charles V and later became Bishop of Lisieux. Oresme used mathematical arguments in his study of physical motion, an approach also taken by several fourteenth-century scholars at the Universities of Oxford and Paris and later by Galileo.

Other Aristotelian philosophers, each in their own way, stressed observation. For instance, the School of Padua—a group of Aristotelian philosophers who, between the fifteenth and seventeenth centuries, tried to define the place of logic in natural studies—also favored empirical investigation. One of them, Jacopo Zabarella (1533–1589), developed a rudimentary method of investigation in natural sciences.

Empiricism, or the lack of it, therefore does not clearly differentiate between "Aristotelianism" and Galileo's "new science." Nevertheless, despite attempts to "mathematize" it, physics before Galileo was largely descriptive and dealt with essences and qualities having little to do with mathematics. Mathematics at that time covered a wide range, including, besides "pure mathematics" (arithmetic and geometry), the so-called "mixed mathematics"—cosmography, geography, hydrology, navigation, meteorology and astronomy, music, civil and military architecture, optics, and other fields that eventually became part of modern mechanics.[19]

By the middle of the sixteenth century most of the surviving classical mathematical texts were known in the West. Old texts had first been translated during the twelfth century, from Arabic to Latin; in the first half of the century, for instance, Adelard of Bath, an English monk who traveled in several Mediterranean countries, translated Euclid's *Elements,* and later Gerard of Cremona, the most industrious translator of the period, translated Ptolemy's *Almagest* and many other texts. At the beginning of the thirteenth century, Leonardo Fibonacci of Pisa wrote his *Liber abbaci,* a treatise on arithmetic and elementary algebra that was instrumental in introducing Indo-Arabic numerals into Europe. By the end of the thirteenth century many other important works had been translated, including those of Archimedes. During the fifteenth and sixteenth centuries many Greek original manuscripts became available, affording Renaissance mathematicians a more complete and precise picture of classical mathematics.

The enormous Renaissance achievements in mathematics were not only translations of classical texts but included original contributions and practical applications, for example, in commerce, medicine, and the arts. Many leading Renaissance mathematicians were not academics but humanists, painters, sculptors, architects, engineers, goldsmiths, stonecutters, and merchants, who used and eventually advanced mathematics. Many studied in abacus schools (special schools for merchants) that existed mainly in the fourteenth century but persisted for some time afterward.[20] One of the greatest Renaissance mathematicians, Niccolò Tartaglia (1499 or 1500–1547) was self-taught and earned his living as an abacus teacher. He spent most of his life in Venice and not only translated, taught, and commented on Euclid and Archimedes but was a codiscoverer of the solution of the third degree equation. This type of discovery was often kept secret, and Tartaglia faced a bitter controversy over priority with Girolamo Cardano (1501–1576), another important sixteenth century mathematician. Tartaglia also made important contributions to military engineering, in particular to gunnery. He discovered that a projectile's trajectory is longest if expelled from a gun elevated at an angle of 45 degrees and designed a square to aid gunners in estimating the range of their guns.

Surprisingly, Galileo never mentioned Tartaglia in his writings, but it seems plausible that Tartaglia did influence him in some way.[21] The work of one of Tartaglia's pupils, Giovanni Battista Benedetti (1530–1590), was strikingly similar to some of Galileo's early efforts.

A specific group of Renaissance mathematicians, the sixteenth century astronomers, merits a place in the history of mathematical progress and is of particular importance to an assessment of Galileo. The best known, of course, was Nicolaus Copernicus (1473–1543), whose work raised an important question, whether astronomy was only mathematics or also physics. Astronomy, in Copernicus's day, was the branch of mathematics that provided the tools to predict phenomena in the heavens (in Greek: *astron,* "star," *nomos,* "law"), mainly for astrological purposes (meteorology dealt with the sublunary region), and still had nothing to do with cosmology or celestial physics.

In his *De revolutionibus orbium coelestium* (1543) Copernicus hypothesized that the earth and the planets revolve around the sun. Whether and to what extent Copernicus was influenced by the works of Aristarchus of Samos—the Greek astronomer who first proposed the heliocentric theory—and whether Copernicus intended his model to be a *true* description of the world, not simply a mathematical device that more accurately predicted celestial events, are topics of wide scholarly debate, which began with publication of his work shortly before or immediately after his death. His editor, Andreas Osiander, without letting Copernicus know, added an anonymous preface presenting the heliocentric theory as no more than a fictitious mathematical instrument. Osiander's classical preface reflects well the separation between a mathematical-hypothetical astronomy and a physical, "true" depiction of the world. It said (emphasis added by present author):

> For it is the duty of an astronomer to compose the history of the celestial motions through careful and expert study. Then he must conceive and devise the causes of these motions or hypotheses about them. *Since he cannot in any way attain to the true causes,* he will adopt whatever *suppositions* enable the motions to be computed correctly from the principles of geometry for the future as well as the past.[22]

Osiander may be considered the founder of instrumentalism, the view that all scientific theories can best be understood as practical instruments for such purposes as calculation and prediction of events, rather than the unveiling of nature's secrets. This judgment was adopted in 1616 by Cardinal Robert Bellarmine (1542–1621), one of the leading Catholic theologians of the time (later saint) as the official view of the Church to oppose Galileo's claim that the Copernican hypothesis could be the true description of the world. Bellarmine thus went beyond the purely epistemological question of the role of scientific theories to include the question of whether Copernicanism could be regarded as a proven truth or just a viable hypothesis.[23] The Galileo affair thus concerned both the truth of nature and the nature of truth.[24] It is interesting to note the shift in the definition of astronomy between the 1612 and 1680 editions of the *Vocabolario*—most probably an outcome of Galileo's trial—toward an emphasis that astronomy is no more than an instrument. The 1680 edition said that astronomy is "the science that treats the course of the skies and the stars, *what we today clearly intend for astrological purposes*" (the words here in italics were added to the earlier edition).[25]

The same problem was faced by another leading Renaissance astronomer, the Dane Tycho Brahe (1546–1601). Tycho worked at first under the patronage of Frederick II of Denmark, who, in 1576, gave him the funds to build an observatory on the island of Hven (Ven), not far from Copenhagen. Later, in 1599, Tycho became a mathematician to the Holy Roman Emperor in Prague. In his Danish observatory, which he called Uraniborg (Castle of the Heavens), Tycho made astronomical observations, shortly before the telescope was invented, with the aid of extremely precise instruments, reaching the extraordinary precision of almost 1/60 of a degree.

Tycho discovered that mere transformation of coordinates could put the center of the system wherever one liked. He admitted that the Copernican proposal to place the sun at the center was mathematically advantageous but for various reasons, both scientific and religious, believed that the earth was at the center. He therefore devised a compromise: The system moves around the sun and the sun around the earth. The mathematics of Tycho's compromise, like Copernicus's system, accounted for the motion of planets with only a few epicycles; its physics, as Kepler noted, destroyed the crystalline spheres. Yet unlike Copernicus's system, Tycho's was adopted in the seventeenth century by the Jesuit astronomers.

Tycho's assistant and successor as imperial mathematician, Johannes Kepler (1571–1630), was more of a theorist and adopted Copernicus's views. Kepler's work in mathematics produced important results; his *Nova stereometria doliorum* (Solid Geometry of Wine Barrels) in 1615, for instance, was a precursor to the infinitesimal calculus, a direction that Galileo's followers also pursued. Kepler's Pythagorean, mathematical, and mystical convictions that planets moved according to simple geometrical relations induced him to look for a mathematical law to express the relationship of the distances between the planets. Although his attempt to fit the five regular solids (tetrahedron, cube, octahedron, dodecahedron, icosahedron) into the five gaps existing between the six known planets did not succeed, it led him to the discovery that Mars's orbit was elliptical rather than circular, and to the formulation of his three laws concerning the motion of planets.

On the basis of this brief outline of the state of scientific knowledge on the eve of Galileo's birth, what should we then call Galileo? Mathematician? Physicist? Philosopher? Astronomer? To call him a "mathematician," in either the old or the modern sense, would be unjust. Galileo himself insisted he was more than a mere mathematician, and "mathematics" in our day has come to have a more restricted meaning than in Galileo's time. Galileo was certainly a first-rate philosopher (in both the old and the modern sense) but, again, this term embraces only part of his contribution. We might perhaps call him a "natural philosopher," but this denotes more the Aristotelian philosophers who occupied themselves with nature, without mathematics. We might call him a "physicist," but this is a modern term, and in any case Galileo's accomplishments were not confined to physics. The same can be said for "astronomer."[26] Galileo's contemporaries and followers, as we shall see, had a similar difficulty in finding an appropriate designation for him because he created new fields that outstripped the terminology of those days. And we have problems today because we still do not fully understand Galileo's contribution to scientific thought. So who was Galileo, and what mark did he leave on science and culture? Let me begin by outlining his life and scientific work.

1

Galileo: The Public Figure

Pre-Telescope Years

The details of Galileo's life and work are generally well known, and the reader will find little new in the outline that follows. However, I present it as a background to the work done in his wake and also highlight certain significant elements, such as his atomistic view, his opinion of the proper place of astronomy in the mathematical and physical sciences, and the importance he attributed to the political and propagandist aspect of his work that influenced the activities and behavior of his followers.

Galileo, like Vesalius, Gilbert, and Kepler, was an academic who became a courtier. As a child he lived in the intellectual atmosphere of the humanists gathered around the Tuscan court; his father, Vincenzio Galilei, was a prominent figure in Renaissance music and collaborated with the humanists and musicians active at court. Later, Galileo was educated at the ancient monastery of Vallombrosa, near Florence, where he was supposed to be initiated into the religious life, and was then sent to the University of Pisa to study medicine. Neither way of life appealed to him; he was more attracted by the geometry of Euclid and the mechanics of Archimedes, which he studied in private with Ostilio Ricci, a mathematician who, like Galileo, was both an academic and a courtier.

Ricci had a mixed cultural background. He had probably been a pupil of Niccolò Tartaglia and was certainly influenced by Leone Battista Alberti (1404–1472; Thomas Settle found among Ricci's manuscripts material copied from Alberti).[1] Ricci tutored the pages of the grand duke and in 1593 became a member of the Accademia del Disegno—the Tuscan official fine-arts academy. Galileo would later perform similar tasks.

The early reports of Galileo's youth are perhaps more legendary than true. Niccolò Gherardini, one of Galileo's earliest biographers, reported in 1654 that when Ricci was tutoring the pages at the Pisan court (Florence was the capital but the court used to spend much time in Pisa), Galileo used to hide in an adjacent room and listen.[2] As we shall see later, this story may well be apocryphal, for the style in those days was to include fictitious anecdotes in the biographies of great men illustrating the casual way in which they came to their future occupations. Yet Gherardini would hardly have reported it if Galileo had not, already as a student, spent considerable time at the Tuscan court. And Ricci's lessons, as Settle relates, were also shared by Giovanni de' Medici, the illegitimate son of Grand Duke Cosimo I. From his early student days, then, Galileo was certainly closely associated with the Tuscan court.

Another early and not very reliable description of Galileo's youth, also dated 1564, was by Vincenzio Viviani, Galileo's young assistant. Viviani said (or fantasized) that in Galileo's third or fourth year in Pisa, around 1583, he had already made his first major discovery in physics, the uniformity of the vibrations of a pendulum. According to Viviani, Galileo came upon the principle by chance, while observing a swinging lamp in the Cathedral of Pisa.[3] Although this report also is unlikely to be true, Galileo must already have acquired a certain fame as a scientist by the end of the decade, since in 1588 he was invited to lecture to the prestigious Florentine Academy about the shape, location, and size of Dante's Hell and a year later began lecturing on mathematics at the University of Pisa. Galileo received 60 florins a year, one of the lowest salaries paid a university lecturer.

Although Galileo at that time had not yet published anything, he left a large number of notes from the period.[4] Two sets were commentaries on questions relating to Aristotle's logic and physics.[5] Another set dealt with Archimedes' hydrostatic principle and motion in a fluid (gas or liquid).[6] Viviani claimed that around 1590 Galileo had already discovered that bodies of different weights fall at the same speed and had performed the famous experiment from the top of the Leaning Tower of Pisa. (He supposedly dropped bodies of varying weights that reached the ground simultaneously, thus publicly disproving Aristotle's statement that the speed of falling bodies is proportional to their weight.)[7] This story too, as will be seen later, is suspected to have been invented.

Both Viviani and Gherardini related that Galileo had a series of disagreements in Pisa, compelling him in 1592 to transfer to a position at the University of Padua. This may well have been true, since Galileo picked quarrels easily. Padua was in the Republic of Venice, an independent and cosmopolitan state. Although the University of Padua had always been one of the leading centers of Aristotelian tradition, the Venetian authorities guaranteed its scholars freedom of thought, and Galileo undoubtedly felt greater latitude there to develop his unorthodox theories in physics and astronomy. In Padua he adhered to Copernicanism, which eventually proved to be both his fortune and his misfortune. He expressed interest in the research fruitfulness of Copernicanism as early as 1597 in two letters, the first to his friend Jacopo Mazzoni, a professor of philosophy in Pisa and Rome, and the second to Kepler.[8]

It is often pointed out that freedom of thought was not at that time Galileo's main worry, and he may have moved to Padua chiefly to improve his financial condition—he was offered an initial salary of 180 Venetian florins. Yet, as Antonio Favaro, the modern editor of Galileo's works, has remarked, his new salary was not much higher than in Pisa, since the Tuscan florin was worth two and a half Venetian florins.[9] Galileo might perhaps have expected it to rise quickly, or hoped to be useful in Venice's important shipyards. Indeed, in Padua Galileo occupied himself with several state projects, for instance, studying the efficiency of oars in naval galleys, and invented a handy machine to pump irrigation water at low cost. He also devised several instruments, one of them a multipurpose "Geometrical and Military Compass," which he produced in his own home for sale.

Galileo remained in Padua eighteen years—the most important period of his scientific work—and had a successful academic career there. But despite steady increases in salary and income from the sale of inventions and instruments and from private tuition, he still badly needed money. He had a family to maintain, two daughters and a son by his mistress, Marina Gamba, and other financial family obligations back in Tuscany. If he could find a position in the service of a wealthy prince, he thought, his financial situation might improve. A first step in this direction came in 1601 when he was indirectly approached by the Tuscan court through his friend Girolamo Mercuriale, a professor of medicine at the University of Pisa, who encouraged him to become the tutor of the eleven-year-old Crown Prince Cosimo. Mercuriale foresaw a future position at court for Galileo and emphasized to him that this contact might one day bring good fortune.[10] Galileo also considered employment by the Duke of Mantua, but did not get the conditions he desired.

The change in Galileo's fortunes came after his great discoveries in physics and astronomy. In 1604 he communicated to the Venetian philosopher and historian, Paolo Sarpi, his most important discovery in physics, the law of free fall. He had discovered that the "spaces traversed in natural motion [i.e., free fall] are in squared proportion of the times [of fall]."[11] Amazingly, Galileo claimed to have derived the law from a premise that the speed of a falling body is proportional to the *distance* covered. This premise is false; the speed, he should have said, is proportional to the *time* of fall. However, Galileo also made a mistake in reasoning that canceled his other substantive error and led him to the right result. Galileo's erroneous deduction has given rise to a debate among historians of science, begun in 1939 by Alexandre Koyré who was the first to use it to try to reconstruct Galileo's heuristic reasoning, an attempt repeated often, and possibly improved, by others.[12]

A year later, in 1605, Galileo began tutoring the Tuscan crown prince during the summer holidays, which he used to spend in Tuscany. He taught Prince Cosimo the use of his compass and dedicated his first book—*Le operazioni del compasso geometrico et militare,* published in 1606—to the prince. The dedication was one of his efforts, following Mercuriale's advice, to cultivate his relations with the Medici.

An important political event in the same year may have been one of the causes of Galileo's decision to leave Padua. Following Venice's stubborn resistance to any interference by the Church in its internal affairs, a grave dispute broke out between the Republic of Venice and Pope Paul V. The Pope excommunicated the Venetian doge (head of the republic) and senate and put the republic under an interdict (debarring all its priests from their functions). Venice reacted by expelling the Jesuits from its territory. The Pope's adviser was the leading Catholic theologian and future saint, the Jesuit Cardinal Robert Bellarmine, while the Venetian government was advised by Galileo's friend, Paolo Sarpi, a friar of the Servite order. This controversy did not directly affect Galileo but certainly interfered with his work; he was no longer allowed to maintain contact with the scientists of the Roman Jesuit College, the most important scientific institution in Italy and one of the best universities in Europe.[13]

Galileo's greatest opportunity came in 1608, after the invention of the telescope in the Netherlands.[14] A year later he constructed his own telescope. He first presented it to the Venetian senate as an instrument for naval purposes, but soon also trained it on the sky and made a series of startling discoveries, for example, that the surface of the moon was irregular, with mountains and valleys, thus refuting the general belief that it was a polished and perfect sphere. He found the Milky Way to be a collection of "innumerable stars grouped together in clusters" and not dense either, as Aristotle had thought. He detected four of Jupiter's moons—named Io, Europa, Ganymede, and Callisio by the contemporary and rival German astronomer Simon Mayr (Marius)—proving that not all heavenly bodies revolve round the earth and refuting the claim that a moving celestial body cannot have other celestial bodies moving around it. Galileo later noticed that at certain times Saturn had an unusual appearance, as if it consisted of three bodies. Neither he nor any other contemporary astronomer was able to explain this phenomenon (only fifty years later, in 1659, did the Dutch astronomer Christiaan Huygens suggest that Saturn was surrounded by a ring, which was verified by some of Galileo's pupils). Galileo also succeeded in seeing the phases of Venus for the first time; the existence of planetary phases was predicted by the Copernican theory, and those of Venus are more easily seen than others.

Galileo published his early celestial observations on the moon, the Milky Way, and Jupiter's moons in 1610 in his well-known *Sidereus nuncius,* a small book written in Latin, the language of philosophers. The word *nuncius,* as Galileo himself noted, means either message or messenger. His intention may have been the first, more modest meaning, but the title was generally translated as *Starry Messenger.*[15] The publication of this thin but extraordinary book marked a climax of discovery and brought him great fame as a scientist. It also marked a turning point in his life. From then on, Galileo practically ceased to be an academic and became a courtier. He also occupied himself less with the doing and more with the politics of science.

Post-Telescope Years

Galileo was applauded in Padua, and his salary was raised to 1,000 florins. But by then, even this high salary hardly satisfied him. On one occasion he remarked: "It is impossible to obtain wages from a republic, however splendid and generous it may be, without having duties attached. For to have anything from the public one must satisfy the public and not any one individual; and so long as I am capable of lecturing and serving, no one in the republic can exempt me from duty while I receive pay. In brief, I can hope to enjoy these benefits only from an absolute ruler."[16] Galileo may have been right, yet as we shall soon see, he needed an absolute ruler not so much as a patron but rather as a springboard to carry out his program.[17]

The ruler was to be the newly crowned Grand Duke Cosimo II de' Medici, Galileo's former pupil. Galileo called Jupiter's moons "Medicean planets" and dedicated the *Sidereus nuncius* to the new grand duke. Then he wrote a letter to Belisario Vinta, Cosimo's first secretary, raising the possibility of returning to Tuscany and expressing his desire to be freed from the teaching assignments that hindered his research. In his letter Galileo said plainly: "Should I come back, I should like his Most Serene Highness to be willing to grant me first of all leisure and comfort so that I may complete my work without having to lecture."[18] He outlined the work he had in mind: two books concerning the system and constitution of the universe, three on local motion, three on mechanics, and other works relating to voice and sound, vision and color, tides, the composition of the continuum, motion of animals, and military science. This was the most extensive research program he ever presented.

Galileo was by then a celebrity and, feeling sure he would be welcome to Grand Duke Cosimo, expressed the wish that the title of Philosopher be added to that of Mathematician, "having," he wrote, "studied philosophy for more years than the months I studied pure mathematics." This was a political request with a propagandistic end: If recognized officially as a philosopher, Galileo would receive formal authority to deal with essences. I do not wish to imply that a mere change of title bestowed the authority to discuss matters not previously open to him, only that he could then finally take his stand on the same level as the Aristotelian philosophers against whom he was arguing. In the following chapters I shall try to show that Galileo's request was also a first formal step toward the inclusion of physics among the mathematical sciences.

Cosimo consented to the request and offered Galileo the post of Chief Mathematician at the University of Pisa and the title of Philosopher to the Grand Duke. However, Galileo was still not satisfied and insisted on the title of Mathematician *and* Philosopher to the Grand Duke, to which Cosimo also agreed. As Philosopher and Mathematician, Galileo believed he had received official acknowledgment that mathematics and philosophy were on a par with and nourished each other. Official acknowledgment did not, of course, signify any new intellectual development in Galileo, but indicated the importance he ascribed to the application of mathematics to natural philosophy, and more specifically to physics.[19]

The agreement of 1610 between Cosimo II and Galileo was mutually beneficial. Cosimo was evidently eager to have Galileo in Tuscany, since the presence of the greatest contemporary scientist certainly enhanced the prestige of the Medici, and Galileo promised to dedicate further books to him. With the arrival of Galileo in Florence, in September 1610, Cosimo II was able not only to pursue the cultural policy of the Medici, especially his grandfather, Cosimo I, but also to present himself worldwide as a patron of the new, emerging science. Galileo also offered to help in projects of practical use to the state or of entertainment to its rulers, writing in his first letter to Vinta: "I have some particular secrets, of use both as application and as curiosity and wonders . . . but they cannot be used, or say rather, applied, except by princes, for they make and sustain wars, build and defend fortresses, and spend for their regal amusement very high sums." Galileo probably thought he could further Cosimo's projects in Tuscany as he had done for those of the Republic of Venice. Cosimo did indeed try to develop the Tuscan economy and agriculture, enlarging its fleet and expanding the city and port of Leghorn. But the scientific or technological importance of Galileo's discoveries played only a secondary role. What seemed more appealing, as Mario Biagioli has recently shown, was the symbolic message of Galileo's astronomical discoveries, which Galileo shrewdly presented as a dynastic emblem attesting to the image of grandeur the Medici had created for themselves.[20] The grand duke could thus emphasize that he intended to give Galileo leisure to complete the ambitious program Galileo had detailed, and, although he formally requested that Galileo be at the court's disposal (this was the only obligation), only occasionally did the court request Galileo's services. For instance, he was consulted on the ever-present problem of floods in Tuscany. Cosimo's heirs, as we shall see, did not always have Cosimo's remarkable understanding of the intellectual needs of the scientists in their service and imposed much more on Galileo's followers, sometimes hindering their scientific research.

Galileo, for his part, received a position of high prestige and a title regarded as more creditable than that of mathematician. His salary of 1,000 Tuscan scudi, paid by the University of Pisa, was two and a half times what he would have received in Padua. Of even more value was the fact that he was not even obliged to teach, while retaining great power in the university and, in general, in the Tuscan cultural establishment. As Chief Mathematician at the University of Pisa, he was in complete charge of the teaching of mathematics and strongly influenced other cultural domains, such as the teaching of philosophy. He did not even have to live in Pisa, which was known for its unpleasant and unhealthy air, and resided in the more agreeable surroundings of Florence. Outside the university he was influential, for instance, in matters concerning the Florentine Academy. Galileo also had other important benefits, indeed almost any of the services the Tuscan court could offer, such as those of Tuscan diplomacy and its postal service.

While establishing close and friendly ties with the Medici, Galileo renewed his relations with the Roman College. This institution excelled in mathematics, and when Galileo went to Rome in 1611 to exhibit his discoveries, he was received by its Jesuits with enthusiasm. This would hardly have been possible had Galileo come from the Republic of Venice.

In Rome, he was also elected to the Lincean Academy, founded in 1603 by Prince Federico Cesi and devoted to the study of nature and mathematics. Galileo was naturally interested in the activities of the academies since they, more than any other institutions, contributed to the spread of knowledge. He had taken part in several Paduan academies and throughout his life persisted in presenting himself as a "Lincean" ("Lynx-eyed"), which undoubtedly raised the prestige of the Lincean Academy. Stillman Drake notes that with Galileo's entry into this academy—which did not normally enroll university people—he distanced himself from the university establishment.[21] This was true as far as teaching was concerned. Galileo stopped teaching but continued, of course, to exert influence on the university.

Moreover, after Galileo moved to the Tuscan court, and deviating from his previously announced plans, he shifted emphasis in his work away from purely technical questions in mechanics toward more general cosmological and cultural issues. The Italian (Marxist) philosopher Ludovico Geymonat even says that Galileo was gradually veering from scientific research to a propaganda campaign in favor of the new science and of Copernicanism in particular.[22] Did Galileo mislead the Tuscan court, exploiting its support to fulfill his scientific—and personal—ambitions? The answer is as complex as Galileo's personality and the way he interacted with the people around him. As Arthur Koestler, the novelist and sometime historian of science, made clear, Galileo was a mixture of a great scientist, a first-rate scholar, and a very ambitious man, who combined scientific talents with writing genius and political acumen in the interest of science and also in his own personal interest. It is therefore difficult to say whether he or the new science could have done without his campaign.[23]

Yet Galileo's life was far from a pure success story. Already in 1611, despite his political ability, or perhaps in reaction to his rapid rise, hostile reactions to his theories started appearing. Shortly after his return to Florence he became involved in a series of annoying, though not dangerous, controversies.

The first important controversy concerned floating bodies and took place partly at the Tuscan court, where several Aristotelian philosophers, led by Lodovico delle Colombe, an erudite Florentine, challenged Galileo's views. The floating or sinking of a body (e.g., ice or wood) in water usually depends on its density. The exception, of course, is metals: Needles, for instance, float on water if placed gently on the surface. Why? According to Aristotle, it was because of their shape. Archimedes said it was because they displace water equal in weight to their own. Galileo sided with Archimedes; Colombe sided with Aristotle. Galileo's explanation was inadequate; Colombe's was wrong. The floating of needles is due to surface tension, unknown at the time. Yet, as Joseph Agassi points out, Galileo's position was far superior, since the Archimedean view not only contradicts but also refutes that of Aristotle.[24]

From the point of view of empiricism, it is interesting to note that Colombe defended his view by means of an *experiment* in which he showed that pieces of ebony of different shapes behaved differently in water. Galileo based his reply, the *Discourse on Floating Bodies,* written in 1612, on Archimedes' postulates and applied *deductive reasoning* from basic principles. Galileo, too, described experiments, one in particular being impressive. He proposed adding copper dust to a wax ball (wax by itself floats on water) and showed that Archimedes's theory precisely predicts the quantity of copper needed to make the wax ball sink. Thus, by gradually and continuously varying the weight of bodies immersed in water, Galileo demonstrated the variation in buoyancy. Galileo deduced the outcome of an experiment on the basis of a hypothetical model; his opponents believed in induction from experiment. For Galileo, experiment was a check; for his opponents it was the source of knowledge. Many authors have emphasized Galileo's empirical work; his treatment of floating bodies is a clear example where his opponents, rather than Galileo, made the more empirical argument.

Another controversy, still annoying though not dangerous, came in 1612 with the Jesuit Christopher Scheiner—a professor of Hebrew and mathematics at the University of Ingolstadt, in Bavaria—and concerned sunspots. Can the sun have alterable qualities, or is it perfect, as Aristotle said? Galileo rightly held that the dark spots on the surface of the sun travel and come and go. Galileo wrongly claimed to be the first to observe the spots.

In the same year, peril made its first appearance, though a rather humble one. Niccolò Lorini, a Dominican professor in Florence whom the grand duke highly esteemed, suggested in a private conversation that Galileo's Copernicanism contradicted the Holy Scriptures. And in 1613 Galileo published his *Letters on the Sunspots* in which, for the first time, he supported the Copernican system in print. Such views could have been interpreted as heretical, and heresy then was no small matter; Giordano Bruno had been burned for heresy thirteen years earlier.[25]

A clamorous debate concerning science and faith occurred at the Tuscan court in Pisa on a morning in December 1613.[26] Benedetto Castelli, Galileo's faithful friend and disciple, had just been appointed Professor of Mathematics at the University of Pisa and was invited to breakfast at court in distinguished company, including the young grand duke, his mother, Maria Christina, and Cosimo Boscaglia, a professor of philosophy at the University of Pisa. Having an illustrious scientist like Galileo or Castelli at the table was considered chic and a source of entertainment for the other guests. On this occasion Boscaglia was too tempted to stick to etiquette; he felt he could challenge Galileo's views with Galileo's disciple as the target without risking ridicule by the famous wit himself. Cornering Castelli would be almost as good as cornering Galileo, and at better odds. The key person at court was Grand Duchess Christina, often described as a bigoted and authoritarian woman (though nothing about her character appeared in Castelli's account of the event), and Boscaglia persuaded her to challenge Castelli on theological grounds and to question the orthodoxy of the claim that the earth moves. Thanks to his knowledge of theology, however, Castelli, a Benedictine friar, succeeded in appeasing the grand duchess. He then let Galileo know of the incident.

Galileo's reaction, a few days later, was his famous *Letter to Benedetto Castelli,* in which he tried to refute theological objections to the Copernican System.[27] Koestler said that Galileo's reaction "was a kind of theological atom bomb, whose radioactive fall-out is still being felt."[28] Galileo's letter touched on the problem of the relationship between theology and science. He agreed that the truth of the Holy Scriptures was beyond doubt, but maintained that their meaning can be questioned, especially in matters of natural philosophy, since they can sometimes be misinterpreted or misrepresented by people not well versed in the subject.

Castelli soon widely distributed the letter, thus initiating the process that twenty years later brought Galileo to trial. A copy even reached Francis Bacon in England.[29]

What was Galileo's purpose in writing his letter? And why did Castelli distribute copies? Who made the copies? Was this done on Galileo's instructions? Available documents do not answer the questions.

One thing, however, is clear: Whether the distribution was ordered by Galileo, or done by Castelli, it was certainly a miscalculation; the letter only increased the opposition to Galileo.

In 1614 Galileo was attacked, this time in public, by another Dominican, Tommaso Caccini. Then, at the beginning of 1615, Lorini sent a copy of Galileo's *Letter to Benedetto Castelli* to Rome, requesting an official reaction against Galileo and his followers. That it took a year for anyone to request an official response is not surprising. Galileo wrote the document as a letter, evidently to stress his position on the subject as a private citizen, so that his action did not call for official response. What presumably warranted such a response, in Lorini's assessment, was Galileo's success in convincing the public, or at least some influential people. Lorini expected that the Inquisition would provide the necessary remedies so that "*parvus error in principio non sit magnus in fine*" ("a small initial error does not end in a great one").[30]

In 1615 Galileo wrote an enlarged version of his letter to Castelli, addressed to the Grand Duchess Christina.[31] A private letter thus became an official document: The Tuscan court Mathematician and Philosopher was now speaking. In his Letter to Christina, Galileo stated, among other things, his view of the distinction between physical propositions that can be conclusively be proved and those that cannot: "Before condemning a physical proposition, one must show that it is not conclusively demonstrated. Furthermore, it is much more reasonable and natural that this be done not by those who hold it to be true, but by those who regard it as false."[32] The fact that Galileo addressed this version of his letter to the grand duchess indicates that he probably thought the challenge to his opponents must be intensified rather than softened. He was in no danger of losing favor with the Medici, they never abandoned him, but opposition to his views was growing in Rome.

In April 1615 Cardinal Bellarmine, at one time an admirer of Galileo, wrote a letter to one of Galileo's supporters, Paolo Antonio Foscarini—a Neapolitan Carmelite monk who had written a work in favor of Copernicanism—unofficially but clearly expressing the official disagreement of the Church with Galileo's attitude to Copernicanism.[33] Bellarmine, commenting on Foscarini's work, adopted Osiander's instrumentalist view and said plainly that the Copernican System could be regarded only as a mathematical hypothesis.

Galileo felt he was losing ground and, in late 1615, traveled to Rome to try and defend his theories. He failed. On February 24, 1616 an official panel of theologians pronounced against his Copernican views, and two days later Galileo received an official notification requesting him to abandon the Copernican view. On March 3, Copernicus's *De revolutionibus* was suspended until such time as necessary corrections were made. Foscarini's book, however, was put on the Index, his publisher was arrested, and Foscarini died suddenly a year later. Stillman Drake hints at a possible connection between Foscarini's death, the listing of his book on the Index, and Bellarmine's instruction of 1616 to Galileo: Poisoning, after all, was common in those days.[34]

Galileo was worried, and for good reason, yet he was received and reassured by the Pope. He also received an affidavit from Bellarmine saying that Galileo had been *notified* only that he ought to abandon the Copernican view and was not officially forced to abjure it, that is, to renounce his opinion on oath; no penitence was imposed. At least part of this special "courtesy" toward Galileo was due to his strong political position, a further indication of the advantage of having been in the service of a patron as powerful as the Grand Duke of Tuscany. Such courtesy would hardly been afforded had Galileo been only a professor of mathematics at the University of Padua or of Pisa.

Galileo returned to Florence; he did not consider himself totally defeated, as ensuing events testified. Before long he was involved in an additional controversy, concerning the three comets that appeared in 1618, this time with the Jesuit mathematician, Orazio Grassi, of the Roman College. Grassi was a talented mathematician and a first-rate architect who had designed and supervised the building of the church of the Roman College. In astronomy he was a Tychoist; relying on the absence of parallax between the observers and the comets in Cologne and in Rome, he declared the comets more distant not only than the moon, but also than the planets. Here one can see both the proper use of observation and a readiness to deviate from Aristotle by an opponent of Galileo, contrary to the widespread myth that all his opponents were dogmatic Aristotelians and hostile to empiricism. Galileo said that comets could be unreal as the rainbow. From a modern point of view, of course, he was in error here.[35]

In 1621, during this controversy, Galileo lost his most important supporter; Cosimo II died at the age of thirty-one. The Medici who ruled after him, however, continued to support Galileo.

The controversy over the comets ended with the publication in 1623 of *Il saggiatore,* (*The Assayer*), an eloquent treatise written in Italian, in which Galileo ridiculed Grassi's views and expressed his own ideas about scientific thought. Galileo wrote plainly that the language of the universe is mathematics, and that man's senses do not sufficiently remedy his lack of understanding. He also conjectured that many perceivable effects, such as light and heat, may be explained by a mechanism of particles. Pietro Redondi, the author of the recent thrilling book on Galileo's trial, claims that Galileo's theory of matter, as presented in *The Assayer,* was the main cause of his future misfortune, since it brought into question the Eucharist, the central Christian sacrament.[36]

In the Eucharist, bread and wine are "transubstantiated" into the body and blood of Christ. This transformation occurs at the level of *substance,* while the *form*—the outward appearance—remains that of bread and wine. This mystery, of course, raises many questions relating to natural philosophy, and especially to the boundaries between it and theology. How is substance transformed? How can it escape being perceived? And to what extent is natural philosophy allowed to deal with these questions?

The interaction between theology and natural philosophy on the issue of the Eucharist, as one might expect, is extremely complex and delicate. It has a long history, concisely presented by Redondi. An attempt to rationalize it, stemming from Aquinas, assumed that substance and form are totally independent. This sort of explanation was accepted by Christian theology. However, to explain appearances such as tastes, colors, and odors in terms of indivisible particles, as Galileo did, implied that substance and form are interdependent, which questioned the very existence of the transubstantiation and hence could be regarded as heretical.

Redondi discovered an anonymous document in the Vatican files denouncing Galileo's theory of matter, as presented in *The Assayer.* Redondi believes that this complaint was presented by none other than Grassi and conjectures that this was the main ground for Galileo's trial. Whether Redondi's conjecture is true or false, his discovery indicates how important and controversial the theory of matter was in Galileo's time. In fact, the main interest of Galileo's followers, as we shall see in the following chapters, appears to have been the character of the physical and mathematical continuum.

Scientific considerations apart, Galileo had insulted Grassi and could no longer rely on the support of the Jesuits. Nevertheless, while Galileo's opponents were increasing, so were his adherents. In Rome his support centered mainly around the Lincean Academy and the cluster of Tuscan cardinals and high prelates, including Giovanni Ciampoli, the Pope's secretary, and Cardinal Maffeo Barberini, the future Pope Urban VIII. Barberini was known for his liberal attitudes and had always been a supporter and friend of Galileo, for example, in the controversy with Lodovico delle Colombe, and, as was the custom of the time, had written an ode to Galileo. In 1624, the year after

Barberini was elected Pope, Galileo went to Rome again to pay his respects. He probably thought that, with Urban's election, the Church's attitude would change, and he hoped to be allowed to hold his views openly.

Galileo and Urban met six times. There is no record of their meetings, but evidently Galileo felt he could now safely write on the Copernican system, for he returned to Florence and began his most famous work, the *Dialogue Concerning the Two Chief World Systems* (here briefly labeled *Dialogue*). This book, which appeared in Florence in 1632, is a masterpiece not only of science and philosophy but also of Italian literature. It took the form of a dialogue between three interlocutors: Salviati, who spoke for Copernicus and Galileo; Simplicio, who advocated Ptolemy and Aristotle; and Sagredo, a "learned layman" (in those days the kind of person often associated with a literary academy) who acted as an arbitrator of sorts, with obvious Copernican leanings. Galileo skillfully combined science, philosophy, and rhetoric into a powerful propagandist composition, diplomatically criticizing Aristotelian cosmology and arguing that the sun is the center of planetary orbits, including that of the earth.

As in the case of the *Letter to Benedetto Castelli,* Galileo appears to have miscalculated the reaction of the Holy Office. Obviously, the Holy Office miscalculated its reaction too, for Galileo was given the *Imprimatur,* official permission to print. The few who had the opportunity to read it soon after publication, including the Jesuits of the Roman College, praised it highly. But sales were soon stopped, and Galileo was summoned to appear before the Roman Inquisition. The young Grand Duke Ferdinand II tried unsuccessfully to intervene in favor of his seventy-year-old Philosopher and Mathematician. Tuscany was suffering a period of rapid political and economic decline; the Medici were too weak to prevent Galileo's trial and certainly could not venture into a perilous dispute, as Venice had done in 1606. This decline, obviously unexpected, was one part, smaller or larger, of the miscalculation. In judging the extent to which intellectual intrigues are more political than intellectual, both then and now, I would say that Tuscany's decline was a major factor in the Holy Office's reversal of attitude toward Galileo.

In 1633 Galileo was tried by the Roman Inquisition and forced to abjure his "error." The trial has left many open questions. The main one concerns a document presented then as evidence that, contrary to Bellarmine's declaration of 1616, Galileo had received a special injunction prohibiting him from discussing Copernicanism at all, as well as from holding or defending it.[37] Another mystery is Galileo's secret meeting with the Commissary General of the Inquisition, Father Vincenzo Maculano da Fiorenzuola, at which Galileo was most probably convinced to abjure his error. Redondi conjectures that Fiorenzuola informed Galileo, upon the Pope's instructions, that he was risking a charge of high heresy because of his theory of matter. Hence, according to Redondi, the Pope may have "saved" Galileo by condemning "only" his Copernican views.[38]

Galileo was condemned, for having rendered himself "vehemently suspected of heresy," to "formal imprisonment in this Holy Office at our pleasure"—life imprisonment—and the *Dialogue* was put on the Index.[39] The sentence, pronounced on June 22, 1633, was long, interesting, and carefully phrased, and was to be sent to all apostolic Nuncios and Inquisitors, "especially the Inquisitor of Florence, who shall read the sentence in full assembly and in the presence of most of those who profess the mathematical arts."[40] Life imprisonment was commuted to house arrest, and Galileo was first transferred to Siena, where he served his sentence as a guest of his friend Cardinal Piccolomini, and later allowed to return to his villa in Arcetri, near Florence, where he remained until his death in 1642.

Galileo's Heritage

We are now approaching the kernel of our study. Galileo's trial naturally ended his political activity, and he spent the last eight years of his life in scientific research. In 1638 he published the book that summarized and crowned his work in physics, the *Two New Sciences,* written, like the *Dialogue,* in the form of a dialogue. The "two new sciences" were strength of materials and kinematics, and Galileo relied heavily on geometry.

Despite political inactivity and house arrest, Galileo retained many social contacts. Many people paid him visits in Arcetri, among them friends, disciples, and admirers—including the grand duke and his brother, who were amateur scientists—and illustrious visitors from abroad. Galileo appears also to have given particular attention to former pupils and his young disciples, following with interest the work of Castelli (appointed Professor of Mathematics in Rome in 1626) and especially of Castelli's students, such as Evangelista Torricelli, Antonio Nardi, and Raffaello Magiotti, about whom we shall learn more in later chapters.

The grand duke, visiting Galileo in 1638, mentioned sixteen-year-old Vincenzio Viviani as a promising young mathematician. Galileo had gone blind by that time and in 1639 invited Viviani to move into his house to assist him, learn from him, and—perhaps—prepare to take his place. Viviani was destined to play a dominant role in the Galilean School, at least in documenting its history.

At the end of 1641, Torricelli, Castelli's most brilliant student and one of the few persons who had read Galileo's *Dialogue* soon after it was published and kept Galileo in touch with current events in Rome, moved from Rome to Arcetri to assist him. He was able to do so for no longer than three months; Galileo died on January 8, 1642.

Did Galileo expect his followers to continue his scientific work? Assuming that he did, which work?

Galileo had completed only part of the research program he had outlined to Belisario Vinta before joining the Tuscan court. His two major works, the *Dialogue* and the *Two New Sciences,* contained most, though not all, of what he had planned to write on world systems, general mechanics, local motion, and the tides. He had not covered military science and had only occasionally dealt with other topics on his list, such as sound, light, the continuity of matter and the motion of animals, but some of his occasional remarks related to these topics, for example, on heat and light in the *Assayer,* were noteworthy. In general, he initiated whole new fields that raised innumerable questions, some explicit, some implicit.

A twentieth-century view of the problems Galileo left open three centuries earlier risks the distortion of hindsight and invites the danger of ancestor worship.[41] Indeed, much of the literature on Galileo, as we shall see in the next chapter, is hagiographic. And excessive eulogy has provoked an opposite trend—equally problematic—excessive criticism, at times without adequate understanding. One way to achieve a judicious balance is to survey the work done immediately after Galileo, primarily by his followers and especially during the period soon after his *Two New Sciences* was published but before it was generally accepted. This is the route I have chosen for this book.

The two main fields Galileo investigated were what we now call astronomy and physics. In post-Galilean astronomy, the whole universe had, literally, to be discovered. Its mapping, with the assistance of the telescope, was an immediate, enormous task, ranging from the moon and the planets to the myriads of newly observed fixed stars. New and precise astronomical tables had to be drawn, to account for Galileo's discoveries such as the satellites of Jupiter, and many new puzzles, such as the strange appearance of Saturn and the nature of comets and sunspots, had to be explained. Other quests, of higher order, concerned the general laws governing the universe, and a possible search for evidence for or against Copernicanism, and the extension of Galileo's developments in mechanics to explain the behavior of celestial bodies. Post-Galilean physics, too, contained many new domains to be investigated, particularly the structure and behavior of matter, and kinematics (motion of bodies), generally regarded as the area of Galileo's primary efforts and major contributions.

Galileo started very early in his scientific career to study the theory and structure of matter; his studies of fluid media are echoed in his *Two New Sciences.* His achievements advanced the understanding of such subjects as buoyancy, vacuums, and air resistance and helped lay the foundations of modern hydraulics, hydrodynamics, and aerodynamics. Many objected to Galileo's claim that a vacuum can exist, and, although he devoted much attention to air resistance, he did not reach a general formulation of how to deal with it, even in his *Two New Sciences.* The book also explored extensively the strength of materials, a "new science" in which everything remained to be discovered.

In his studies of matter, Galileo disregarded alchemical theories, some still in fashion in his day, and analyzed material behavior in terms of mechanical effects among atoms. This approach led him to ponder the physical and mathematical aspects of the divisibility of matter, with its deep philosophical implications. The physical question was: Is matter infinitely divisible, that is, continuous, as Aristotle had believed, or divisible only to ultimate particles? Although Galileo expressed atomistic views on many occasions, and was well aware that a solution of the problem would entail the existence of a vacuum, he did not treat the subject systematically in any of his writings and left it unresolved for his followers. The mathematical question seems to have been even more difficult for him, which he frankly admitted, and also was left open. Yet, did the question involve only physics and mathematics, or, as Redondi claims, was it also theological? If it had no dangerous theological implications, would not Galileo have written the treatise on continuum he had promised Belisario Vinta? I will return to this point in detail in chapter 4, on the theory of the indivisible.

In kinematics, too, Galileo left many loose ends such as the formulation of a principle of inertia. Galileo's idea of inertia was not entirely clear and coherent. He spoke of two types: one "terrestrial" and rectilinear, and one "celestial" and rotational. Yet his whole philosophy was based on a Copernican abolition of the celestial-terrestrial distinction. But this, again, is how we see it today, after Newton. Galileo and his followers may have seen it differently.

Galileo left open other questions, perhaps smaller in range but no less important. By perfecting optical instruments he contributed to optics but did not produce any systematic theory. He also investigated the behavior of magnets, with much enthusiasm, but did not attempt theoretical account.

Galileo's scientific work also held the promise of many practical applications, both civil and military. His astronomical discoveries, for instance, were relevant to navigation and the perfecting of the calendar. His achievements in physics were applicable to gunnery and the development of such devices as an air pump. Galileo himself was an instrument maker, as his "Geometrical and Military Compass" and the various telescopes and microscopes he constructed amply testify.

The above are only a few examples of questions left open by Galileo and of new possibilities generated by his science. To what extent did he expect his followers to pursue these avenues and to further his explorations? Although he did not leave specific instructions, his followers did complete many of the projects he either initiated or expressed a wish to carry out but did not. In particular, the continuity of matter and the motion of animals were studied by two of his leading followers, Cavalieri and Borelli, respectively. Developments by other followers will be described in later chapters.

In general, the picture of Galileo's Italian followers is rather gloomy. The problems he touched on naturally occupied many intellectuals not associated with him, and during the second half of the seventeenth century, Italy ceased to be the main center of the scientific revolution. The focus shifted to the other side of the Alps. What happened to Italian science in this period? What were the contributions of Galileo's followers? Can such a review teach us something new about Galileo? Will it require us to reassess current views of his legacy? What caused the "decline" of Galilean science?

I will attempt to answer these questions by studying some aspects of the work of Galileo's major followers, such as Cavalieri, Torricelli, Viviani, and Borelli. These scientists would hardly have been able to pursue his work without a clear idea of his methodology. But Galileo left no outline of his approach to scientific investigation, so the question of his methodology remains one of the basic issues in Galilean studies and one of the subjects of the coming chapter.

2

Galileo's Image through the Ages

The Galileo Myth

I will now break the narrative to deal with historiography. Galileo has often been presented as a legendary, brave, free-thinking scientist, punished for having dared to go against the current, to doubt received wisdom, and to describe nature as he saw it without philosophical or religious bias. He was no doubt an outstanding scientist, but the literature often attributes to him things he never did, thus distorting his true image.

The Galileo myth is multifold and has varied through the ages, exhibiting such contradictory colorations as hagiographic and anti-hagiographic, traditionalist-continuist and revolutionary, "empiricist" and "apriorist." A common theme, for instance, is the conflict between science and religion—at least between science and Catholicism—and depicts Galileo as a martyr and his opponents, chiefly the Church, as obscurantists who hindered the progress of science. Another facet of the Galileo myth, based on a certain assessment of modern science as essentially empirical and detached from prejudices, contends that Galileo, as one of the earliest modern scientists, grounded—and perhaps even founded—experimental science. According to this view, Galileo was an empiricist who, unlike the natural philosophers of his day, sought to answer questions not by reading philosophy books but rather through direct contact with nature. All these interpretations, of course, need not be mutually exclusive.

Myths nevertheless beget iconoclasts and may even give rise to countermyths. Arthur Koestler, for instance, considered Galileo's arrogance, rather than the "obscurantism" of his opponents, to be the source of his misfortunes, and Alexandre Koyré claims that Galileo's work was much less empirical than it may seem and that Galileo may have deceived his readers by describing experiments he never performed.[1]

Where does the myth—or the antimyth—begin, and where does the truth end? Many nuances must be considered. The myth of Galileo as a martyr of science is now receiving considerable attention. In 1979 Pope John Paul II raised the question of the Galileo affair, encouraging new investigations, which, we can hope, will shed new light and allow a more judicious presentation—to use the expression of Maurice Finocchiaro, whose recent work greatly contributed to dispelling the Galileo myth.[2] We will return to Galileo's "martyrdom" in chapter 8 in an account of the Accademia del Cimento. Here we will concentrate more on Galileo as a purported empiricist (and its "a priorist" counterpart), a central problem in Galilean studies. How was this image created, and how has it evolved to the present day?

Galileo's Image as an Empiricist Scientist

The origins of Galileo's image as a scientist who derived his knowledge from sense experience can be traced as far back as the writings of people who knew and worked with him. In 1654, twelve years after Galileo's death, Prince Leopold de' Medici, the grand duke's brother and a patron of Tuscan science, asked Vincenzio Viviani and Niccolò Gherardini, who had had direct contact with Galileo, each to write a sketch of Galileo's life. Viviani—"Galileo's last disciple," as he used to call himself—was by then Galileo's leading follower, and Gherardini, who had assisted Galileo in his old age, was the curate of the *prioria* of Santa Margherita a Montici, in the neighborhood of Galileo's house. Their reports were to serve as a basis for a broader biography the prince planned to include in the first collection of Galileo's works, then being edited in Bologna.[3] This biography has not been found and may never have been written. Viviani's and Gherardini's outlines, however, did survive, though they appeared only posthumously, in 1717 and 1780, respectively.[4]

Viviani's essay, "*Racconto istorico della vita del Sig.r Galileo Galilei,*" was short and eloquent (fewer than forty pages in the National Edition of Galileo's writings). It summarized Galileo's approach to science by stating:

> He said that the letters in which it was written were the propositions, figures and conclusions in geometry, by means of which alone was it possible to penetrate any of the infinite mysteries of nature. He therefore owned very few books, though the best and of the first class. True,

> he praised the good things that had been written in philosophy and in geometry to elucidate and awaken the mind to their own order of thinking and maybe higher, *but* he said that the main entrance to the very rich treasure of natural philosophy was observations and experiments, which, through the senses as keys, could reach the most noble and curious intellects.[5]

Viviani's essay should, of course, be considered in its historiographic context, as a piece of contemporary prose (see chapter 7). As far as Galileo's approach to science is concerned, Viviani's narrative, like Galileo's prose, can be interpreted in more than one way. Viviani may have intended a balanced description of Galileo's methods, as both empirical observation and theoretical speculation, but in effect the empiricist part predominated, at least in later literature. Indeed, this imbalance in the literature is understandable, since Viviani supplied impressive anecdotes to illustrate Galileo's ability to grasp the secrets of nature by observing it with great care down to the minutest detail. He related that an oscillating lamp in the Cathedral of Pisa inspired Galileo to construct his own pendulum and fathom its behavior: "By very precise experiments he verified the equality of its vibrations."[6] He went on to tell how Galileo then applied the principle he had discovered to a series of inventions for measuring time and motion and for medical purposes. Viviani similarly reported (or fabricated) a description of Galileo dropping weights from the Leaning Tower of Pisa to refute Aristotle's theory that bodies fall at speeds proportional to their weight by showing that they fall at equal speeds, "demonstrating this by *repeated experiments* performed from the height of the bell tower of Pisa in the presence of other lecturers and philosophers and all the student body."[7]

How important are Galileo's pendulum observations and Leaning Tower experiments, true or alleged? They are certainly more important to his image than to his science; without them he might have obtained the same results by other means, but his empiricist image could hardly have survived.

Viviani's anecdotes were not confirmed in Gherardini's outline, which otherwise was similar to Viviani's. Gherardini wrote explicitly that Galileo combined speculation and observation, theory and practice.[8] He, too, said Galileo had very few books, and "his studies depended on continuous observation while making deductions from all things he saw, heard or touched, subject[s] of his philosophy. He used to say that the book from which one ought to learn is the book of nature, which is open to everybody."

Since Viviani became a prominent and influential scientist—he practically replaced Galileo at the Tuscan court—his biography was the primary source for Galileo's later biographers. And, as Lane Cooper has shown in his *Aristotle, Galileo and the Tower of Pisa,* later authors portraying Galileo as an empiricist relied on Viviani, sometimes adding imaginary details. Since Gherardini was less well known, completely overshadowed by Viviani, his biographical sketch of Galileo mostly served only to reinforce Viviani's; the historiographic tradition (Galileo as empiricist) can be said to stem largely from Viviani.

The empirical image of Galileo, although not universal in later historical writings, was especially evident during the eighteenth and nineteenth centuries. The Galilean bibliography has become voluminous (more than 2,000 items by the end of the nineteenth century and over 3,000 additional books and articles in the twentieth) and cannot be easily summarized.[9] It exhibits various interpretations and viewpoints, often reflecting the periods in which the items were written, and represents different uses of Viviani's and Gherardini's early sketches.

Some important pioneering works relating to Galileo were written in Tuscany during the second half of the eighteenth century after the Enlightenment, for example, one in 1780 by Giovanni Targioni Tozzetti, a Florentine naturalist and historian.[10] Targioni Tozzetti gave one of the most comprehensive accounts still in existence of seventeenth-century science in Tuscany. It focused on Medici patronage of science, rather than the details of Galileo's work, and thus can be regarded as a forerunner pointing toward the sociology of science. Although Targioni Tozzetti did not contribute directly to Galileo's empiricist image, he did so indirectly by publishing Gherardini's brief biography for the first time.

A more detailed life of Galileo was written several years later (1793) by the Florentine historian Giovanni Battista Clemente Nelli.[11] Nelli's book took on special significance and authority because he had discovered (by chance) and purchased Galileo's and Viviani's papers between 1739 and 1750 (a large part of the "Galilean collection" now in the National Library of Florence) and had at his disposal an impressive collection of sources.[12] However, as Nelli's own manuscripts and drafts (also part of the Galilean collection) testify, he seems to have made little use of Galileo's papers and, in essence, merely amplified Viviani's sketch.[13]

So did many other writings of the eighteenth and nineteenth centuries. For example, Alexander von Humboldt, in 1846, embellished Viviani's story about the discovery of the pendulum with a report that Galileo used a pendulum to measure the height of the Pisan cathedral, a report later (1883) called a "gross mistake" by Antonio Favaro.[14]

Nevertheless, as Galilean studies developed, historians more and more suspected that Viviani and Gherardini had distorted the figure of Galileo and were unreliable sources for the purposes of the modern history of science. In 1887 Favaro claimed to have found among Galileo's unpublished papers evidence that Viviani had changed Galileo's date of birth from February 15 to February 19, 1564 in order to bring it closer to the day Michelangelo died, in Rome, on February 18, 1564.[15]

Approximately a century ago, new documents and studies began to appear that enabled a reevaluation of the Galileo legend. In particular, between 1890 and 1909, Favaro published the National Edition of Galileo's works, the most important and reliable source for research on Galileo. Scholarly publications on the beginnings of modern science gave a better understanding of Galileo's cultural and social surroundings, and developments in the philosophy of science illuminated the nature and role of empirical investigation.

In 1903 Emil Wohlwill, one of the earliest modern Galilean scholars, pointed out that the absence of evidence for the validity of various details in Viviani's story rendered it unreliable by the standards of the modern history of science.[16] Wohlwill doubted, among other things, the story about the lamp in the Cathedral of Pisa, and evidently with good reason, since Favaro later indicated that the so-called Galilean lamp had not been hung in the Cathedral of Pisa until 1587, four years after Galileo was supposed to have watched it swing. Also, although there may have been an earlier lamp, the story remains in doubt because Galileo mentioned the constant frequency of pendulum movement for the first time fifteen years later, in a letter to his teacher Guidobaldo del Monte.[17] No evidence has been found that he knew the principle any earlier, or that he invented an instrument using the principle to count pulse rate. On the contrary, he and his contemporaries were accustomed to using a normal pulse as the best measure of time.

Wohlwill also doubted the story of the Leaning Tower experiment, citing Galileo's failure to mention it in any of his writings and the absence of any other confirming evidence. Moreover, Galileo's early manuscript on motion, published for the first time only in the middle of the nineteenth century (also in the first volume of the National Edition) and known today as *De motu,* contradicted Viviani in some respects. For example, it said that bodies do not fall at the same speed, contrary to the tower test results alleged by Viviani, and Wohlwill believed that it was written during the period in which Galileo supposedly performed the experiment.[18]

Only a few historians shared Wohlwill's doubts; most argued against or disregarded them. Favaro, for instance, believed Viviani reliable and belittled his apparent inaccuracies. Favaro was justified to some extent, since it was common in Viviani's day to add imaginary anecdotes to biographies of important people. However, many historians repeated Viviani's and Gherardini's anecdotes as if they were true, sometimes even embroidering them with invented details.[19] Other historians discounted Viviani but still more or less credited Galileo with a pioneering empiricism. Martha Ornstein, for instance, in her classical work on scientific societies of the seventeenth century, said that "the Jesuits Riccioli and Grimaldi were skillful experimenters, and—a veritable triumph for Galileo's methods—tried to refute him by experiment." (We consider Riccioli's and Grimaldi's experimental work in the next section.) Evidently Ornstein, like many other historians, was taking it for granted that Galileo had founded experimental science.[20]

Another important study, written between 1919 and 1927 by Leonardo Olschki, focused mainly on Galileo's nonacademic surroundings. Olschki drew attention to the self-evident influence on Galileo of the traditions of Renaissance literature, art, and technology. Well aware of Viviani's inexactitudes, but very much in Viviani's and Gherardini's spirit, Olschki said: "Galileo, like all other molders of Italian prose, was a humanist. In his scientific research, he linked empirical knowledge with a rediscovered and thoroughly comprehended blend of Platonic thought and Archimedean method, to their mutual refinement and enrichment. Similarly, he combined in a literary way the norms of national prose with the vital norms of the mother tongue, again to their mutual advantage and enrichment."[21]

Olschki was certainly right in emphasizing Galileo's literary background. The existence of so many literary academies then, in some of which Galileo actively participated, shows the keen interest of the learned public in literature. Yet Galileo's nonacademic background is still generally uncharted territory. We know very little about such aspects as the relations between the science-based professions and arts and Galileo's work. Fields such as civil engineering, architecture, sculpture, and painting relied on geometry, but how much did men of mathematics like Galileo contribute to them and vice versa?[22]

The influence of Renaissance philosophy and Galileo's academic affiliations on his science has received scholarly scrutiny. J. H. Randall and A. C. Crombie, for instance, suggested that during the eighteen years (1592–1610) Galileo spent at the University of Padua he might have been exposed to the empirical school that developed there between the fifteenth and seventeenth centuries, when a number of Aristotelian Paduan philosophers devoted themselves to searching for fruitful ways to serve natural philosophy.[23] This enterprise culminated in the work of Jacopo Zabarella (1533–1589), a philosopher and logician, who proposed a rudimentary method of investigation based on the notion of "regress" (*regressus*) whereby, in a series of stages, knowledge induced first from observation is then reapplied to yield deeper understanding.[24] Zabarella also was empirical in his work and described observations of meteorological effects. Randall and Crombie believed it reasonable that Galileo, who on one occasion praised Zabarella, may have been influenced by the Paduan school, but the extent of such influence on Galileo's thought and extensive use of mathematics is difficult to estimate.[25]

More recently, Charles Schmitt, the late historian of Renaissance culture, emphasized a difference between Galileo's empirical method and Zabarella's regressus.[26] With Zabarella, said Schmitt, an observation comes before a conclusion, so the conclusion is a posteriori; Galileo's experiments, even those apparently performed before his Paduan period, were designed a priori as embodiments of preconceived theories. Thus, Zabarella might be called an a posteriori empiricist, who tended to form his ideas after observation, and Galileo an a priori experimentalist, whose ideas tended to precede observation, which makes the role of experiment less important for Galileo than for at least some Aristotelian philosophers. However, the distinction may be less rigorous than it appears, for in practice neither theoretical conclusions nor observational data always necessarily come first.

Galileo's empirical methods—whether a priori or a posteriori—were also emphasized in recent studies. Thomas Settle, for instance, tried to reconstruct the sequence of events that brought Galileo to his most important discovery in physics, the law of free fall, by analyzing his *De motu,* presumably written between 1589 and 1592, and his *Two New Sciences,* published in 1638.[27] Settle concluded that Galileo probably alternated theory and experiment in a somewhat trial-and-error procedure yielding an unexpected result. From what Galileo said in his *De motu,* Settle thought it likely that, in his early days in Pisa, Galileo experimented by throwing objects from a tower (not necessarily the Leaning Tower) to investigate the relation between "impressed force," that is, imposed on a body from outside, the body's weight, and the body's motion. On another occasion, while investigating uniform motion, he experimented with bodies rolling down inclined planes, which led him to the discovery of the law of uniform acceleration. Galileo later reported his conclusions from these experiments in a systematic way in his *Two New Sciences.*

If this reconstruction is true, then experiment was, indeed, vital to Galileo's work and thought, and the results might not have come without experiment. The question nevertheless remains: Galileo may have been an *experimenter,* that is, performed experiments; was he also an *experimentalist,* that is, did he regard experiments as the source or merely the proof of scientific discovery?

Galileo's empiricism was considered further by Stillman Drake in an analysis of Galileo's manuscripts. Drake examined volume 72 of the Galilean manuscript collection, containing notes, diagrams, and calculations made between 1602 and 1610 in Padua. Some had never been published, even by Favaro, because they consisted of calculations and diagrams alone with no illuminating propositions or explanations. Yet Drake believed that they held significant clues to the way Galileo arrived at some of his discoveries. On the basis of his proposed chronological order and interpretations, Drake reconstructed the sequence of Galileo's experiments, mainly those related to free fall. The details of Drake's interpretation and the debate that followed are too complex for a brief summary, but they left little doubt that Galileo was a very active experimenter. Drake also believed that Viviani's biography of Galileo was reliable and that Galileo did perform the Leaning Tower experiment. Furthermore, according to Drake, Galileo may also have used prior experiment as a basis for subsequent theory.[28]

Nevertheless, the fact that Galileo did not mention the work described in his manuscripts in any of his published writings implies that he may have encountered serious problems, possibly due to self-admitted (but not declared) inexact results, and was reluctant to make them public. It is difficult, if not impossible, on the basis of the historical records to come to a general conclusion as to Galileo's attitude toward these dutifully noted but not published experiments.[29]

One element, then, is common to many writers on Galileo: an emphasis on the empiricist character of modern science in general and of Galilean science in particular. Was Galileo's science as empirical as they said, or only as they expected it to be?

If by empiricist we mean one who observes nature or performs experiments, then Galileo was surely an empiricist. But then so were many of his predecessors, for example, Leonardo da Vinci and Vesalius. Empiricism in this sense was not peculiar to Galilean science. Also, as shown in the previous chapter, Lodovico delle Colombe, Galileo's opponent, tried to refute Galileo's theory of floating bodies by means of experiments. However, by a stricter definition an empiricist is not merely one who observes nature or performs experiments, but rather one who professes these observations and experiments to be the primary source of knowledge. But even if Galileo met this definition, here, too, he was preceded by many Renaissance philosophers, such as the Paduan school. Moreover, some seventeenth-century scientists who said that science did not need empirical demonstrations nevertheless performed them. René Descartes, for instance, believed that truth could be grasped only by means of innate, a priori ideas but devoted much of his time to empirical observations, such as anatomical dissection, justifying them by saying that only God has complete wisdom, whereas man is compelled sometimes to have recourse to experiment.[30] Where then, does Galileo stand?

We have already cited many claims that Galileo's empiricism, unlike that of his predecessors, relied intensively on experimenting, primarily to confirm preconceived ideas, so that he was an a priori empiricist. Thus, by definition, the importance of observation and experiment was relatively diminished. But what role did Galileo himself assign to experiment? Did he use it as a basis for theory construction? As a final test? Or only as a didactic demonstration? These questions cannot be answered on the basis of either his work or his manuscripts.

I propose to explore whether Galileo's contemporaries and followers can help supply the answers.

Galileo's Contemporary Critics

Until fairly recently, historians of science often judged the correctness of a theory largely by whether they found themselves in agreement with it. For instance, one who agreed with a theory of Galileo paid little attention to an opposing "incorrect" theory. Lately, historians have tried to evaluate theories more in context, so "incorrect" theories are also taken into consideration.[31] Increasing attention is now given to doubts expressed by Galileo's contemporaries concerning the truth about some experiments he claimed to have performed.[32]

One of the first, if not the first, to express such doubts was the French philosopher Marin Mersenne, as early as five years before Galileo's death. In his work *Harmonie universelle,* published in 1636, Mersenne complained that he repeated certain experiments described by Galileo but got different results, suggesting that Galileo may not have done them. For instance, Mersenne compared the rolling of a sphere down an inclined plane with the falling of a sphere dropped vertically. He was probably testing one of Galileo's statements in the First Day of his *Dialogue*.[33] After he failed to duplicate Galileo's reported result, Mersenne concluded, "I doubt whether Galileo did perform the experiments of falling on a plane, since he says nothing about them, and since the propositions he gives often contradict the experiment."[34]

The discrepancy between Mersenne's and Galileo's results, as Thomas Settle noted, was clearly caused by differences in the effect of the rotational inertia of their spheres.[35] Since the concept of rotational inertia did not appear until Isaac Newton's time, many years later, and neither Galileo nor Mersenne gave evidence of knowing of it, the experiment was for them effectively unrepeatable. One can hardly blame Galileo for being the source of the error; in the *Dialogue* he described no experimental details. He did so only in the *Two New Sciences,* and there he described an experiment different from Mersenne's. Galileo compared the times required by a rotating sphere to travel different distances along a plane, Mersenne the distances traveled by a rotating and a falling sphere in the same time.[36] In modern terms, the linear distance s traveled in time t by a sphere rolling down a plane inclined at an angle θ to the horizontal is given by the formula $s = \frac{1}{2}gt^2(5/7)\sin\theta$, where the factor 5/7 accounts for the moment of inertia and is necessitated by the rotation. In the falling case, without rolling, $s = \frac{1}{2}gt^2$.[37] If one compares, as Galileo did, the distances s_1 and s_2 traveled by a rolling sphere in times t_1 and t_2, and expresses the comparison as a ratio, the factor $\frac{1}{2}g(5/7)\sin\theta$ cancels, leaving simply $s_1/s_2 = t_1^2/t_2^2$, which is not affected by the moment of inertia. This ratio could not be obtained in Mersenne's case, so his results remained affected by the moment of inertia. Yet Mersenne was not satisfied even after the *Two New Sciences* was published. In 1645, not long after Galileo's death, he wrote several letters to Torricelli asking for clarification of Galileo's inclined plane experiments (see chapter 5).

Mersenne expressed other doubts. Later in *Harmonie universelle,* for instance, he described other experiments with falling bodies that yielded results much smaller than the one Galileo reported in the *Dialogue*.[38] And three years later, in 1639, in *Les nouvelles pensées de Galilée* (pp. 72–73), Mersenne argued (among other disagreements) against Galileo's claim that pendulums of equal length are isochronous (have the same period of oscillation).

Mersenne was not the only contemporary to question Galileo's pendulum experiments. In a series of experiments performed in 1640, the Jesuit scientist Giambattista Riccioli, aided by Francesco Maria Grimaldi, tried unsuccessfully to make a pendulum with a period of exactly one second and was not able to account for his failure.[39] Today we know that simple pendulums of a given length are not exactly isochronous; they are approximately so only for small angles of swing no greater than five or six degrees. Galileo may indeed have gone too far in his assertions. But his opponents on more than one occasion misunderstood him and performed inappropriate experiments.[40]

Riccioli was anti-Galileo. Mersenne, like Sagredo in Galileo's *Dialogue,* was a "neutral" observer. Yet even some of Galileo's supporters, for instance, Antonio Nardi in Rome, voiced doubts about his alleged empiricism. During the 1640s Nardi wrote a long outline—nearly 1,400 pages—of the scientific knowledge in his day under the title *Scene* (*Scenes*). Regrettably, his interesting manuscript (now in the National Library of Florence) was never published.[41]

In the *Scene* Nardi explained why Mersenne's inclined-plane experiment did not succeed.

> I am certain that in experiments with a ball sliding in a groove, its motion is a composition of two motions; while it descends it tumbles, because of the load and adherence to the groove; because of the load and adherence, this [combined motion] is slower than falling motion in air.[42]

Nardi was right that both translation motion along the plane, and rotation of the ball, caused by friction, must be taken into account. It is remarkable that Nardi gave the correct physical explanation many years before the notion of moment of inertia was introduced.

Nardi here defended Galileo's theory. However, on the same page of his manuscript, in a discussion of Galileo's theory of projectiles and in words very similar to Mersenne's, he questioned, on empirical grounds, Galileo's law of free fall, one of the components of the motion of projectiles:

> Galileo's science of motion of falling bodies and projectiles relies on two principles: one, that the horizontal motion is equal, the other that the motion of falling bodies increases in proportion to time. Yet I doubt that this second principle agrees with experiments, hence the speed of falling does not follow this principle.[43]

If Galileo's law of free fall was inexact, so of course was his theory of projectiles.

Nardi was not the only follower of Galileo to question the empirical grounds of the projectile theory. Giovanni Battista Renieri, a gunner who attempted to apply the theory to the aiming of artillery, complained in 1647 to Torricelli that his guns did not behave according to Galileo's predictions (see chapter 5 for details). Torricelli replied—

contradicting the traditional belief that Galileo and his followers were thoroughgoing empiricists—that Galileo spoke the language of geometry and was not bound by any empirical result.

The remarks of Mersenne, Riccioli, Nardi, Renieri, and Torricelli referred to single experiments, or at the most a narrow range of topics, and did not dispute Galileo's work in general. But some modern historians have used them (sometimes out of context) to construct an antihagiographic image of Galileo. The first to do so, at the end of the nineteenth century, was Raffaello Caverni, a Florentine priest who questioned the entire traditional image of Galileo and created the "anti-Galileo myth."

The Galileo Anti-Myth

Between 1891 and 1900, Raffaello Caverni wrote a monumental work (six volumes), relying on many previously unpublished documents, that may be regarded as the first broad modern study of Galileo. He attempted to relate the place and importance of Galileo's work to that of his predecessors, his followers, and other seventeenth century European scientists. The result was a totally new representation. As the title of Caverni's work, *Storia del metodo sperimentale in Italia* (*History of the Experimental Method in Italy*), testifies, he believed that the new science created during the seventeenth century was indeed experimental. However, Caverni contested the received tradition, claiming that Galileo had been given too much credit at the expense of earlier Renaissance scientists and contemporaries. He boldly proposed that Galileo had only pretended to make many of his supposed discoveries, and went so far as to denounce him for claiming priority for the accomplishments of many other mathematicians, such predecessors as Tartaglia and Stevin and even Galileo's own followers, for example, Castelli and Cavalieri. Biographers, said Caverni, had reported and occasionally amplified Galileo's false claims.

Caverni's work has many merits. Most importantly, he drew attention to the work of many scientists whom Galileo had overshadowed. He also pinpointed one of Galileo's greatest contributions to science: recognition of the basic value of geometry in physics. Admittedly, Caverni let his "anti-Galileanism" carry him too far, and some of his claims were exaggerated. He failed to see Galileo's achievements in the synthesis of previously unrelated scientific fragments and disregarded the fact that seventeenth century standards of credit and acknowledgement differed from modern criteria.[44] Nevertheless, Caverni made a significant contribution to the history of science and is still a useful source for Galilean historians.

Caverni's views encountered much opposition, at least in Italy. Not only had he dared to question the accepted Galileo myth, but he did so shortly after the unification of Italy, when Galileo attained the status of national hero.[45] Caverni's work received strong criticism, he himself was isolated (excluded from the board of editors of the National Edition), and his findings were to a great extent ignored. The printing of his sixth volume was even interrupted when he died in 1900, although he left a complete manuscript. Only recently was it reprinted, when historians came to realize its importance.[46]

Despite opposition, Caverni's work had an impact, even in his own day, though initially mostly outside Italy. It marked a turning point in Galilean studies and encouraged a critical review of Galileo's work. Before Caverni, to the best of my knowledge, no one had investigated Galileo's intellectual predecessors—indeed, it was taken for granted that Galileo had none.[47] A few years after Caverni's work appeared, Pierre Duhem published his studies of Galileo's predecessors, emphasizing Galileo's debt to them.[48] Neither Caverni nor Duhem directly challenged Viviani's account; Caverni even praised Viviani.[49] Yet the implication of discredit was picked up in 1939 by Alexandre Koyré in his *Galileo Studies,* which initiated a general assault on Galileo's traditional empiricist image.

Koyré opened a new era in Galilean studies with a reinterpretation of Galileo's work. Koyré, rejecting the accepted view of Galileo as an empiricist, yet without belittling him as Caverni and Duhem had ("Every historian, and especially every biographer, is something of a hagiographer," Koyré mused), conjectured that neither experience nor experiment had played an essential role in Galileo's work. He even suggested that some experiments Galileo described in detail had never been performed, basing his proposition both on evidence and even more on his view of Galileo's methodology.[50]

Koyré frankly presented his conclusion as an outcome not so much of his remarkable study of the Galilean heritage as of his own view of the role of experiment in science. An experiment, he maintained, is a question put before nature, and the resulting "facts" have to be ordered, interpreted, and explained within the language in which the question is formulated; thus the results are unavoidably constrained and a priori. This interpretation, claimed Koyré, is particularly appropriate when applied to the birth of modern science, when mathematical language had evolved further than experimental ability. (This will be clearly seen in chapter 5, dealing with the case of ballistics.)

Also, an obstacle in the way of experimental science up to the end of the seventeenth century had been the lack of instruments for precisely measuring time. Galileo's method, rather than his specific exploration of the physical world, fits this interpretation, his method having much in common with the Platonic and, more specifically the Archimedean view, of scientific procedure.[51]

According to Plato, the changing world of experience can be understood only in terms of ideal forms, and forms are geometrical figures. Thus, by reference to forms, he considered the theory of planetary uniform motion in a circle to be intellectually satisfactory and self-explanatory. The forms can be grasped intuitively a priori and described by indubitable first principles or axioms of deductive science. Euclid and Archimedes, each in his own way, went one step further and constructed an axiomatic—deductive and abstract—system of mathematical physics. It is debatable whether or not Galileo was in fact so heavily committed to the concept of ideal structure. According to Koyré, Galileo's great achievement was going one step beyond Plato and Archimedes in granting movement to the abstract and immovable Archimedean bodies. The laws of his physics were thus deduced abstractly, without recourse to experiment on real bodies. Therefore, said Koyré, the experiments Galileo claimed to have performed, even the ones he really carried out, could not be anything but thought experiments. This does not mean Koyré denied that Galileo did empirical work, only that the empiricism in his experiments was based on a priori conceptions, rather than experiential, and hence was also relatively unimportant.[52]

Koyré's views have aroused a lengthy debate among historians of science. Some, such as Winifred Wisan, tend to minimize the role of experiment in Galileo's work, saying that "the formal role of experiment, for Galileo, was not to confirm principles indirectly but to render them immediately evident."[53] Others, such as Thomas Settle, Stillman Drake, and James MacLachlan, reject Koyré's view, arguing that he makes blanket claims about Galileo's lack of use of experiments and too frequently takes an a priori approach to history without paying attention to facts.[54]

Nevertheless, Koyré's main claim remains unrefuted, despite the weakness of its detailed description. He maintained that, since all experiments are premeditated, whether or not Galileo performed them is not important for understanding his intentions. Since this was what Koyré thought, he did not care about Galileo's experiments, apart from proving that they were not in any way the source of Galileo's ideas. Koyré's opponents may thus have failed to understand that he was more concerned with Galileo's *methodology* rather than details of his *method.*[55] One may also point out that Koyré's contention is difficult to rebut because it concerns what Galileo should have done rather than what he actually did.[56]

There is, then, still no consensus among historians on the methodological aspects of Galileo's contributions to physics and especially to scientific method. So far the question appears unanswerable on the basis of the evidence. Perhaps it is even anachronistic since, after all, it is a modern philosophical problem that may not be translatable into seventeenth century terms.

In my assessment, both current views on the methodological aspect of Galileo's work—empiricist versus a priorist—suffer from the same weakness. Both ascribe to Galileo traits that best fit the holders' own view of science and expect him to be consistent and unproblematic, to exhibit a clear and straightforward methodology and to behave accordingly. But as Finocchiaro has recently shown, Galileo, unlike some other scientists and philosophers, did not leave us an unequivocal, systematic account of his methodology and his attitude toward experiment.[57] For instance, Francis Bacon, in his *Novum organum,* declared all science to be based on experiment and all a priori reasoning to be faulty. To the contrary, René Descartes held that truth can be grasped only by means of innate ideas, and Immanuel Kant, in the eighteenth century, said clearly that science must be a priori.

Galileo left a variety of material, with many scattered methodological notes that can often be interpreted in more than one way. Typical perhaps is a famous remark in the *Assayer:*

> Philosophy is written in this grand book, the universe, which stands continually open to our gaze. But the book cannot be understood unless one first learns to comprehend the language and read the letters in which it is composed. It is written in the language of mathematics, and its characters are triangles, circles, and other geometric figures without which it is humanly impossible to understand a single word of it; without these, one wanders about in a dark labyrinth.[58]

Despite its apparent clarity—geometry is the language of science and therefore fundamental for scientific research—the comment is far from easy to interpret, and Galileo barely developed it any further. Neither his published works nor his manuscripts give a precise idea of the role he assigned to observation and experiment or the relation he saw between them and mathematics. Furthermore, his approach to his work may have changed in the course of his long life, and so perhaps did his attitude toward the use of experiment in physical research.[59] Finally, Galileo may have not been entirely consistent in what he said or thought, that is, his method may not have resulted directly from his methodology, or vice versa.

However, the issue is not only Galileo's "consistency" but also the validity of the historian's or philosopher's reconstruction of his methodology. Historians who expect simple and universally applicable rules will selectively find support for their preconceptions and interpret as inconsistency whatever happens not to fit; the problem then is with the interpreters' hermeneutical ideal.[60]

Thus, the question of Galileo's empiricism remains open. The current discussion of Galileo's experimental activity is, on the whole, a debate between holders of modern views of science citing historical illustrations rather than an inquiry into Galileo's thought. In the process, contesting parties have expounded the case history of Galileo at great length but, except for Caverni, have paid little attention to the activities of the followers among whom Galileo lived and worked. In Galileo's wide correspondence are letters from pupils, collaborators, learned friends, and admirers. If their interest were more historical and less philosophical, the contestants might possibly have given greater consideration to direct witnesses of Galileo's work, since presumably the activities of these witnesses, being at least partially inspired by Galileo, might teach us something more about his.

I propose therefore to examine the work of Galileo's followers. I do not intend to suggest that this can resolve the questions about Galileo; We have already cited conflicting opinions. Viviani and Gherardini may have presented the empiricist aspects of Galileo's efforts, while Mersenne, Riccioli, and Nardi questioned his empirical work, and Torricelli rejected it. Galileo's followers may have included unreliable witnesses, and one of the theses of this examination is that some of them were indeed unreliable. Also, it is not clear to what extent those regarded as Galileo's followers actually shared his thought. Nevertheless, Viviani did write his biography of Galileo at least partly on the background of the Galilean school, and the life and work of Galileo's followers may help us to understand why Viviani portrayed his teacher the way he did.

There were also occasions when Galileo's followers genuinely represented their teacher, mainly when called on to defend his theories. On other occasions some of his followers, for example, Cavalieri and Borelli (to be discussed in the following chapters), went beyond representation and introduced elaborations or extensions of their own. Expressions that originated with Galileo's followers, whether or not they were acting as his spokesmen, were inspired by him and may be regarded, with precautions, as an indirect heritage from him. I call "Galileanism" knowledge that Galileo at least indirectly inspired and that can in part be attributed to him.

Who were Galileo's followers? The next chapter gives an outline of the life and work of scientists who tried to continue Galileo's scientific work.

3

Galileo's "Followers"

The Galilean School

What is a "follower"? Even Viviani, who was certainly well informed about the people connected in various ways to Galileo and compiled a list of the leading figures, had difficulty in finding an appropriate word for them. In one of his handwritten drafts he vacillated between *scolari* (pupils), *seguaci* (followers), and *discepoli* (disciples).[1] However we define the term, followers were important in history of science; without them the contributions of any innovative scientist could hardly spread. Newton's ideas, for instance, were disseminated, at various levels, through the help of his many followers including the circle of closely associated "Newtonians" that included disciples such as John Keill and Colin Maclaurin. Other followers who spread the word were more loosely associated with Newton, for example, Edmund Halley, who encouraged Newton's efforts and helped him publish his *Principia,* Richard Bentley, who, in his "Boyle Lectures" delivered in 1692, argued that Newton's theories proved the existence of God, and even Voltaire, who helped to popularize Newton's theories on the Continent.[2] Galileo's similarly varied followers regarded themselves as part of a movement, which they frequently described as a "school."

Many historians, including Galileo's early biographers, have emphasized the importance of such scientific schools; for instance, not only Viviani but Targioni Tozzetti and Nelli (in the eighteenth century) listed many people whose work was related to Galileo's.[3] Also, Antonio Favaro, in volume 20 of his National Edition, helpfully supplied short biographies of some of them.[4] But "movement" and "school" are not sharply defined. Which "followers," at what level of devotion, qualify? Must a follower agree with the master and, if so, to what extent?

The vagueness of the term "follower" and conflicting views as to

Galileo's scientific legacy make it difficult, perhaps even impossible, to decide who were Galileo's followers. Historians have so designated assorted types of intellectuals, sometimes even the anatomist Marcello Malpighi and the poet John Milton. One can, of course, find some tenuous connection; Malpighi, for instance, was briefly a pupil of Borelli, who undoubtedly was a Galileo follower, but does this automatically enroll him in the Galilean school? Perhaps the only thing Malpighi and Galileo had in common was the fact that Malpighi used an instrument Galileo had improved—the microscope. As for Milton, he may have met Galileo in 1638 during a visit to Italy (no reliable evidence confirms this) and did later mention Galileo in his writings.[5]

Thus, although some historians claim Galileo created a cultural movement that spread his ideas—implying possibly hundreds of followers who "agreed" with him—others deny the existence of a Galilean school and cite many differences between Galileo and his direct collaborators.[6]

I do not intend to discuss here the social boundaries of Galileo's science. However, although his followers may not actually have constituted a school or a movement, his wide correspondence (we know of more than 4,200 letters) does indicate that many intellectuals, in Italy and abroad, declared their admiration for his ideas and helped to disseminate them, even when—as in the cases of Mersenne and Nardi—they sometimes disagreed with them. Thus, if we include all Galileo's collaborators, pupils, dozens of learned friends, and many admirers, we can say that he no doubt had a large number of followers. But if we restrict ourselves to intellectuals who shared his thoughts and concerns, or even to those who extended his work, the number becomes much smaller.

Despite such a diminished number, and the strong opposition to Galileo's ideas (also the generally low reputation of mathematics in Galileo's day), Galileo's immediate followers were highly respected and obtained chairs of mathematics or other important positions all over Italy. By the middle of the 1630s, Castelli occupied the chair of Rome, Cavalieri the prestigious chair of Bologna, and Borelli that of Messina. The chair of Pisa was in the hands of Galileo's followers, even though not always the famous ones, throughout the century; in 1626, vacated by Castelli, it passed to Niccolò Aggiunti, then, in 1636, to Dino Peri, and finally, from 1640 to 1647, to Vincenzio Renieri, an Olivetan monk who collaborated with Galileo. The reader may never have heard some of these names, an indication of how little is still known of Galileo's direct collaborators. Galileo, then, seems to have had considerable influence in appointments to chairs in Italian universities. Exactly how he exercised his influence may be impossible to reconstruct, because he left no direct evidence. One thing is clear: Galileo's followers acted as missionaries for his ideas and kept him well informed of how they went about it.[7]

Who were Galileo's direct collaborators and what was their contribution to science? Can their works and their relations with Galileo tell us anything new about him and his science, or help us understand why science declined in Italy during the second half of the seventeenth century? The following is a brief account of the lives and activities of Galileo's foremost collaborators, mainly those who worked with him in his old age, when most of his ideas had reached a final formulation.

Historians agree that Galileo's most faithful follower was Benedetto Castelli (1578–1643), a pupil, co-worker, friend, and supporter who defended and disseminated Galileo's ideas. Born in Brescia in northern Italy, Castelli was a Benedictine friar who studied under Galileo in Padua and later collaborated with him in many projects, making contributions to hydraulics, optics, and astronomy. In 1613, on Galileo's recommendation, Castelli became a professor of mathematics at the University of Pisa. One of his students there was Bonaventura Cavalieri, later one of the greatest seventeenth-century mathematicians. Castelli introduced him to Galileo, and Cavalieri became one of Galileo's leading followers.

Although Castelli devoted much time and energy to teaching, like Galileo he also maintained good relations with several Italian courts. At the Tuscan court he was chiefly an adviser in matters of hydraulics, winning the admiration and friendship of the grand duke; he also tutored the grand duke's brother, Lorenzo, and was often seen in the company of the Medici. In 1626 Pope Urban VIII asked him to come to Rome to tutor the Pope's nephew Taddeo Barberini. Castelli moved to Rome, where he was also appointed papal consultant on hydraulics and professor of mathematics at the University of Rome.

Castelli's published work, his correspondence with Galileo, and indirect evidence indicate how active he was as both a scientist and promoter of Galilean science. In his major scientific efforts he extended Galileo's theories on floating bodies and laid the foundations of modern hydraulics; in 1628 he published his most important work, *Della misura dell'acque correnti.*[8] Equally important was Castelli's promotion of Galileo's ideas among Roman intellectuals, which is our main interest here.

Rome during Castelli's years there, especially before Galileo's trial, bustled with scientific activity. In the vanguard of the participants were the outstanding Jesuit scholars of the Roman College, the lively members of the Lincean Academy, and a group of pupils Castelli had gathered around him and called "The Galilean School," which included the excellent students Evangelista Torricelli, Giovanni Alfonso Borelli, Michelangelo Ricci, Raffaello Magiotti, and Antonio Nardi. Castelli mentioned this "school" in a 1641 letter to Galileo introducing Torricelli. "You will see," wrote Castelli, "how much honor he will bring to your great school" ("Alla gran scola di V. S. Ecc.ma").[9]

Torricelli—the best known of these young scientists—was born in 1608 near Faenza, in Romagna, in the northern part of the Papal States. He studied mathematics and philosophy at the Jesuit school in Faenza, excelled in his studies, and was sent by his uncle to Rome to study under Castelli. He became Castelli's assistant, and in 1632, during Castelli's absence from Rome, he kept Galileo informed on the reception of his *Dialogue* in Rome and expressed adherence to Copernicanism.[10] Castelli proved to be right in foreseeing Torricelli's future as an outstanding and successful scientist. For a time Torricelli was secretary to Ciampoli, Galileo's influential friend and supporter, who may have taught him the political skills that were later to help him further his career. In 1641 he moved to Florence to assist Galileo himself and replaced him as Court Mathematician after Galileo's death the following year. More details on Torricelli's work at the Tuscan Court will come later in this chapter.

Another outstanding pupil of Castelli was Borelli, who was Torricelli's age and came from the south of Italy (probably born in Naples). While in Rome Borelli may have been in touch with Tommaso Campanella (see Introduction note 9). Castelli had a high opinion of Borelli, who became a lecturer in mathematics at the Messina studium in 1635 on Castelli's recommendation and, after Torricelli's death, was internationally regarded as the leading Italian scientist.

Borelli's work was broad and complex, ranging over geometry, physics, astronomy, hydraulics, life sciences, and even vulcanology. Possibly deterred by such complexity, historians of science have done little research on Borelli, except for Koyré who investigated Borelli's astronomy despite great difficulty in reading his writings.

Borelli's earliest scientific work, published in Messina in 1646, dealt with mathematics and was a polemical response to two controversial papers on geometry written two years before by a Palermitan mathematician named Pietro Emmanuele. A year later, when Messina was hit by an epidemic of fever, Borelli displayed his insight into epidemiology and wrote a study of the possible causes of the disease. In 1656 he moved to Tuscany where he remained for the most productive decade of his career. He held the chair of mathematics at the University of Pisa, did research in anatomy and astronomy, and took a leading part in the work of the Accademia del Cimento.

Unlike Borelli and Torricelli, the other three members of Castelli's Galilean School, Ricci, Magiotti, and Nardi, were nonacademics and served the Papal court. They remained faithful to Galileo after his trial and after his death collaborated extensively with Cavalieri, Torricelli, and other mathematicians all over Europe.

Michelangelo Ricci (1619–1682) was a wealthy Roman intellectual. Despite the duties imposed by a series of high-ranking positions at the Papal court, he persisted in his mathematical studies and came to be regarded, after Cavalieri's and Torricelli's deaths, as one of Italy's leading mathematicians. Like Mersenne in Paris, he corresponded with many contemporary mathematicians in Italy and abroad, keeping Galileo's followers informed about new mathematical results. Though never ordained, he was made a cardinal of the Church in 1681.

Raffaello Magiotti (1597–1658) was a scriptor (copier of manuscripts) of the Vatican library. (Caverni lamented that Magiotti's talents were suffocated in library dust.[11]) In 1638 Castelli proposed him for the chair of mathematics in Pisa, but Magiotti chose to stay in Rome where he continued collaborating with Torricelli and Galileo's other followers all over Italy.[12]

Very little is known about Antonio Nardi, not even the dates of his birth and death. Yet Antonio Favaro, editor of Galileo's manuscripts, remarked that "few among the Italian scientists in Galileo's day deserve to be as accurately studied as Nardi does."[13] Nardi's work holds particular interest for the present purpose. Like most other followers of Galileo, he wrote on many topics, but—for as yet unknown reasons—none of his works were published, perhaps explaining why he does not appear in the *Dictionary of Scientific Biography.*[14] Yet, although Nardi may not exist in today's official history of science, the correspondence of Galileo's followers shows he was highly esteemed as a scientist by his contemporaries. Copies of his manuscripts, including the *Scene,* were sent to Galileo's Tuscan followers and are now kept in various Florentine libraries. Nardi's *Scene* will be cited as a useful source, so a brief description is in order.

Nardi initially had in mind a work on geometry, but the *Scene* evolved to include other subjects (in 1645 Ricci wrote to Torricelli: "Nardi . . . has completed the metaphysical part; now he works on physics and later will revise the mathematics").[15]

Like Galileo's *Dialogue,* Nardi's *Scene* was written in Italian, probably for the "educated layman." It was 1,392 pages long, divided into nine *scene* (scenes) and 282 *vedute* ("views," in the sense of chapters). It dealt mainly with geometry but also such fields as astronomy and cosmography, physics and chemistry, metaphysics, theology, philology, and even moral considerations. As Nardi himself admitted, the *Scene* may at first sight appear confused,[16] wandering from one topic to another, for instance, abandoning a mathematical discussion to engage moral or theological problems. However, a table of contents shows up toward the middle of the manuscripts and orders the text according to the various subjects.[17] Despite this unusual arrangement, the *Scene* seems nearly ready for publication and is easy to read, thanks to the clear handwriting and Italian style.

Maurizio Torrini, co-editor of the correspondence of Galileo's followers, remarked that Nardi's *Scene* "seem to be an indiscriminate review of contemporary culture, in which the new science drowns, rather than emerges."[18] In my opinion, it is precisely this aspect that makes Nardi's work so important: It presents science in the *context* of contemporaneity, as seen by a Galilean intellectual soon after Galileo's death.

Galileo had many other supporters in Rome besides Castelli and his Galilean School. These included eminent intellectuals such as Giovanni Ciampoli (1589– or 1590–1643), secretary to Urban VIII and member of the Lincean Academy, who encouraged Galileo to write and publish his *Dialogue;* Virginio Cesarini (1596–1624), secretary to Pope Gregory XV, chamberlain to Pope Urban VIII, and member of the Lincean Academy, to whom Galileo addressed the *Assayer;* and Gasparo Berti (1600–1643), a highly esteemed mathematician and physicist who, after Castelli's death in 1643, took his place at the University of Rome (a short time before Berti himself died).[19] The list can be extended almost indefinitely, depending on the interpretation of "followers," but is here restricted to close followers of Galileo who shared his thoughts and concerns and extended his work. And Caverni, one of the best-informed historians on Galileo's followers, commented that "Immediately after Castelli, one should consider, in this splendid Senate of Italian science, Bonaventura Cavalieri."[20] In fact, while Castelli was training a young generation of Galilean scientists in Rome, Cavalieri, in the north of the country, was developing the "theory of indivisibles," which would attract the particular interest of these scientists. Who, then, was Cavalieri, and how was he related to Galileo and Galileo's other followers?

Cavalieri

Cavalieri, like Torricelli, was a pupil of Castelli whom Castelli introduced to Galileo. He spent most of his life north of the Apennines and collaborated with other Galileo followers through a regular and massive correspondence. He shared Galileo's thoughts and concerns and worked on one of the major problems Galileo left unsolved: geometrical (and implicitly also physical) continuity. His main contribution was a general mathematical method permitting the relatively easy calculation of areas and volumes. Galileo's leading followers studied, discussed, developed, and also criticized Cavalieri's work more than any other topic. Cavalieri also developed further and promulgated other aspects of Galileo's work, such as the theory of projectiles.[21]

Cavalieri, like Galileo and many other Renaissance scientists and artists, was immortalized in a misleading biography, by one of his pupils, Urbano d'Aviso, published, together with a treatise by Cavalieri on the sphere, in 1682, thirty-five years after Cavalieri's death.[22] D'Aviso stated, for instance, that Cavalieri was born in Milan in 1598, a date considered uncertain.

In 1615, according to indirect evidence, Cavalieri was received into a minor order of the Jesuats (*gesuati*) (founded in the fourteenth century and devoted to caring for the infirm, not to be confused with the Jesuits). Active all over the peninsula, it adhered, in Cavalieri's day, to the rule of St. Augustine. (D'Aviso was also a Jesuat.) Cavalieri was sent to the Pisan Jesuat monastery of St. Girolamo where he met Castelli, then living there, and Castelli initiated him into mathematics and introduced him to Galileo. Cavalieri, as Galileo himself reported in one of his letters, excelled in his studies, learning unaided the works of Euclid, Archimedes, Ptolemy, Apollonius, and other mathematicians.[23] In 1617 Cavalieri was in Florence and a year later was asked to substitute for Castelli in teaching mathematics at the University of Pisa. In 1620 he returned to northern Italy and served his order in Milan, Lodi, and Parma. It was during this period that he developed his theory of indivisibles, hoping to get a position in a university with Galileo's help.

D'Aviso claimed that Cavalieri was highly esteemed by Galileo as his favorite pupil. Today we know from Cavalieri's correspondence that good relations did not always flow both ways between the two scientists, but the custom in biographies at that time was to emphasize or invent such happy ties. Although Cavalieri was always faithful to Galileo, Galileo explicitly refused to acknowledge him as his pupil and,

Figure 3.1. Bonaventura Cavalieri. (Collezione Gioviana in the Museo di Storia della Scienza, Florence)

in 1632, even accused Cavalieri of having plagiarized some of his discoveries.

The accusation was an episode in a long history of complex relations between the two scientists. In 1619 Cavalieri asked in vain for Galileo's help in obtaining the chair of mathematics at the University of Bologna, after its incumbent, Giovanni Antonio Magini, died. It is not particularly surprising that Galileo did not support Cavalieri on this occasion; after all, Cavalieri was still quite young, Magini's chair was very prestigious, and Galileo himself had been refused the chair when he was older than Cavalieri. It is, however, astonishing that Galileo sometimes failed to help Cavalieri on later occasions.

When Castelli left Pisa in 1626, Cavalieri applied for the vacant chair, but by the time his application reached Tuscany, a less qualified candidate, Niccolò Aggiunti—a philosopher and former pupil of Cavalieri—had been appointed. The grand duke appointed Aggiunti, according to Favaro, following Galileo's strong recommendation.[24]

Why was Aggiunti chosen so quickly and Cavalieri disregarded? Why was the grand duke so eager to have Aggiunti in Pisa? If Galileo supported Aggiunti (which he must have done, as chief mathematician at the University of Pisa), was it because of the grand duke's wishes, or perhaps because Aggiunti was a more "faithful" and "reliable" follower? Available documents offer no answer to these questions.

It was only in 1629, when the chair of Bologna was again vacant and Cavalieri applied for it, that Galileo recommended him warmly, comparing him to Archimedes. But even then, as seen from correspondence between Galileo, Cavalieri, and Cesare Marsili (a learned Bolognese gentleman, friend of Galileo, and Cavalieri supporter) Galileo's recommendation was not obtained easily. When Marsili requested Galileo to affirm that Cavalieri had been his pupil (such affirmation must have been indispensable, since Marsili underlined it), Galileo refused. Why?[25]

The whole relationship between the two scientists was rather strange. Galileo archives contain 112 letters from Cavalieri, written between 1619 and 1641, relating to his work and asking for advice. Only two letters from Galileo to Cavalieri survived, which does not prove that more were not written, but from Cavalieri's letters it is clear that Galileo did not always answer. Although Galileo may not have agreed with Cavalieri's theories, his attitude also seems to have been calculated: Cavalieri, despite his enormous talent, had no particular political pull, so writing to him was not likely to bring Galileo much benefit. As Koestler noted, Galileo treated Kepler in much the same way.[26]

With or without Galileo's encouragement, Cavalieri's work steadily progressed. In Bologna he taught, in a three-year cycle, Euclid's *Elements,* the theory of planets, and Ptolemy's *Alamagest.* His first printed work, the *Directorium generale uranometricum* (1632) dealt with trigonometry and with logarithms and their application to plane and spherical trigonometry and astronomy, and contained the first logarithmic tables published in Italy.

Also in 1632, Cavalieri published his well-known *Lo specchio ustorio* (The Burning Glass), a small volume on conic sections and their applications. This book, like Galileo's *Dialogue,* was a popular work, written in Italian, that attempted to show how useful mathematics can be.

Lo specchio ustorio caused a small scandal between Galileo and Cavalieri. It included, among other things, a treatment of the motion of projectiles, and Galileo, as soon as he had read it, complained to Marsili that Cavalieri had plagiarized the theory from him. Cavalieri wrote Galileo a long letter of apology that was a masterpiece of courtesy compared to the rudeness of Galileo's complaint. Cavalieri explained that he had acknowledged only a general debt to Galileo, without entering into details, since he had not seen all of Galileo's works and was not sure that Galileo had not changed his opinions. He declared that he was ready to make any necessary changes in the book, would even burn all copies, and was stopping its distribution until he received an answer from Galileo. Galileo cooled down, accepted Cavalieri's apology, and even thanked him for having mentioned his work.[27]

A year later, Galileo was put on trial and humiliated. He could no longer afford to spurn the friendship of Cavalieri, who was by then one of the foremost mathematicians of the time and had always been a faithful adherent. It was probably in this period that Galileo finally decided to acknowledge Cavalieri as one of his leading disciples. In 1635 he offered him the chair of Pisa, but this time Cavalieri chose to remain in Bologna.

That same year Cavalieri published his major work, *Geometria indivisibilibus continuorum nova quadam ratione promota* (Geometry Advanced in a New Way by the Indivisibles of the Continua). This was a long book (over 500 pages), written in Latin and with many obscure passages. What makes it particularly difficult to read is Cavalieri's avoidance of algebra in favor of descriptive geometrical demonstrations, as was common in his day. Yet his *Geometria* remains a major classic in mathematics (to be discussed in the next chapter).

Cavalieri's work, in general, was monumental, no fewer than eleven books on mathematics. He also presented many of his results in correspondence with Galileo and other contemporary scientists, such as Torricelli in Italy and Beaugrand and Mersenne in France. Cavalieri's theory of indivisibles was challenged by Paul Guldin, a Jesuit mathematician, in his *De centro gravitatis,* also known as *Centrobaryca,* published between 1635 and 1641 in Vienna. Cavalieri replied, in 1647, with a book entitled *Exercitationes geometricae sex* (Six Geometric Exercises). Although it did not succeed in refuting all Guldin's claims, *Exercitationes* is important for a general understanding of Cavalieri's theory. It was Cavalieri's last work; he died November 30, 1647.

So much for Cavalieri's life and relations with Galileo. We now return to what was happening in Tuscany, the main center of Galilean science.

Galileo's Tuscan Followers

Although some Galileo pupils were making his science known all over Italy, his major sphere of influence remained Tuscany. Galileo and his followers were always present in all the major Tuscan intellectual centers, such as the University of Pisa, the leading Florentine academies, and, of course, the Tuscan court.

As mentioned above, Galileo always made sure that the chair of mathematics in Pisa was filled by one of his followers, but he did not always select the best-qualified candidate. When Aggiunti died in 1635, and Cavalieri declined Galileo's offer to take his place, Castelli proposed Borelli. Galileo, for the second time in a decade, seems to have preferred a less-able candidate, Dino Peri, a close friend of Aggiunti.[28] Peri was the first Galilean follower, to the best of my knowledge, to use the term "school" for Galileo's adherents. In a 1633 letter to Galileo, who was then in custody in Siena, Peri wished Galileo's school success all over Italy, despite persecution.[29] Was Peri's appointment a reward for his devotion? Who knows what scientific advances might have resulted from direct collaboration between Galileo and Borelli? Borelli did finally receive the chair of Pisa, in 1656, fourteen years after Galileo's death.

Galileo's followers were also active in all the leading Florentine academies, such as the Accademia della Crusca, the Accademia del Disegno, and in particular the Florentine Academy. In 1621 Galileo was elected consul of the Florentine Academy, a position previously held by one of his assistants, Mario Guiducci.

These and other Galileo followers collaborated with him and among themselves and helped to spread Galilean science in Italy and abroad. It was also thanks to their help that Galileo was able to complete and publish the *Two New Sciences*, in 1638. Nevertheless, it was Castelli, either directly or his pupils, who was most instrumental in this mission perhaps because from 1610 until his trial Galileo does not seem personally to have trained any students of his own. Most of Galileo's Tuscan followers in this period were intellectuals who started collaborating with him at a fairly advanced stage of their careers; they were not "formed" by him. Not that they were incapable of acting as his spokesmen, but as a matter of fact they were not the Galilean "missionaries" that Castelli and his group were.

Luckily, some promising pupils were brought to Galileo during the last years of his life. In 1639, Grand Duke Ferdinand II introduced the young Vincenzio Viviani to Galileo, and Galileo later invited Viviani into his house as his amanuensis. In 1641, on a visit to Galileo (by now blind) in Arcetri, Castelli suggested that Torricelli be asked to come to Tuscany to assist Galileo in writing two additional "Days" of his *Two New Sciences*. Galileo agreed, and in October 1641 Torricelli joined Galileo and Viviani in Arcetri. Torricelli and Viviani became friends and later collaborated in many works, including, in 1644, the famous barometric experiment that bears Torricelli's name.

Thus, when Galileo died, he left only two young close co-workers to carry on his researches, Torricelli and Viviani. This accounts for Viviani's trouble in finding an exact name for Galileo's followers. In an early draft of his life of Galileo, he may have had in mind those who studied under Galileo, and so chose *scolari* (pupils).[30] When he realized how few they were, in a second draft he switched to "followers," then finally "disciples."[31] Had Galileo devoted more time to his students he might have created a generation of disciples rather than only a few to broadcast and advance his ideas more effectively.

Here, we will consider primarily the work of Galileo's Tuscan followers, under the patronage of the Tuscan court. Torricelli was the most important.

At the Tuscan Court: Torricelli, Viviani, Borelli

Only a few general histories of science pay attention to Tuscany and Italy as a whole during the period that followed Galileo's death, as if his disappearance from the scene ended Tuscan science.[32] Yet the record shows intense scientific activity in Tuscany for at least twenty-five more years, mostly at the Tuscan court, with significant results, especially during the five years Torricelli was in the grand duke's service.

In 1642, shortly after Galileo's death, Grand Duke Ferdinand II appointed Torricelli court mathematician in Galileo's place. Torricelli was invited to live at court and was appointed lecturer of mathematics at the Florentine Academy with an annual salary of 200 scudi—not as high as Galileo's but certainly more than Torricelli expected.[33] Viviani related that the grand duke reestablished for Torricelli the chair of mathematics at the Florentine studium (as explained in the Introduction, the Florentine Academy almost replaced the studium of Florence; the more important chair of mathematics at the University of Pisa was then occupied by Vincenzio Renieri).[34] Torricelli also became a member (*innominato*) of the literary Accademia della Crusca and, according to Caverni, in 1642 created the experimental Medicean Academy, for the purpose of applying Galileo's discoveries, though I have found no other evidence of the existence of this academy.[35] At the beginning of 1644, Torricelli was appointed lecturer on military fortifications at the Accademia del Disegno, the Tuscan official fine-arts academy, with a yearly salary of 40 scudi.[36] The choice of Torricelli, for court scientist as well as the other positions, was fortunate; he was an excellent successor to Galileo and, despite the difficulties that emerged after Galileo's trial, knew how to preserve, extend, and promote Galileo's science. But his formal position at court had some curious aspects.

First, Torricelli's title. A seemingly negligible semantic detail, it cannot be overlooked in the light of Galileo's strong insistence on the title of Court Philosopher and Mathematician. It is, however, not at all clear what Torricelli's title was; reports are contradictory. In his life of Galileo, Viviani said that Torricelli inherited the "glorious title of Philosopher and Mathematician" from Galileo; Torricelli's coffin—he was buried in the Church of St. Lorenzo in Florence—reportedly bears the inscription "Magni ducis Etruriae Mathematicus et Philosophus" (The Mathematician and Philosopher of the Grand Duke of Tuscany); and the same title appears in both Torricelli's will and a note added to the will in 1662 by a registrar named Ulisse Magnani.[37] However, in 1672 Viviani also wrote, in a short biography of Torricelli, that Torricelli was the "Mathematician of His Highness" (*Matematico di S.A.*) but not philosopher.[38] Torricelli was also called the Grand Duke's Mathematician (*Magni Ducis Mathematico*) on the inner front page of his *Opera geometrica* (1644), the only work published in his lifetime, and Cavalieri referred to Torricelli in his letters as "Court Mathematician," not Court Philosopher.[39]

I have tried in vain to find a written agreement between Ferdinand II and Torricelli defining Torricelli's position at court. Since most of Torricelli's Tuscany manuscripts and papers survived, it is unlikely that such an important official paper—if it ever existed—got lost.[40] The only document we have relating to Torricelli's work at court is his appointment as lecturer at the Accademia del Disegno.[41]

What, then, was Torricelli's title?

Figure 3.2. Evangelista Torricelli. (Collezione Gioviana in the Museo di Storia della Scienza, Florence)

Paolo Galluzzi, the coeditor of Torricelli's correspondence, takes it for granted that the title was simply Court Mathematician and suggests that the grand duke may have withheld the title of Court Philosopher in a pretense of separating mathematics from philosophy to accommodate the instrumentalist ruling of the Church that a mathematical theory should not describe a physical reality.[42] This, if true, points to the growing caution of the Medici in dealing with science after Galileo's trial. Since Torricelli's *Opera geometrica* was printed by the court and widely distributed, it would have been prudent to designate the author as merely the "court mathematician." The coffin, being buried, could safely be inscribed "Mathematician *and* Philosopher." Actually, we do not know if Torricelli even wanted or ever asked for the double title.

It seems more likely that a formal agreement between Ferdinand and Torricelli never existed, that the agreement was tacit, and that Torricelli's title remained undefined. The issue seems much less important in the case of Torricelli than it was in the case of Galileo, who was hypersensitive about the trappings of prestige, and perhaps we should not judge by Galileo's standards. The Tuscan court had previously had quite a few mathematicians in its service whose main task was tutoring pages. Galileo himself began his work at court as tutor of the crown prince, although he skillfully attached special significance to the title, which he then used for political purposes. Successor scientists like Torricelli and Viviani were much less "politicians" and may not have insisted on the more impressive title. Also, Torricelli may have felt less urgency to formalize a link between mathematics and physics. In the aftermath of Galileo's work, and despite his trial, it was clear all over Europe that mathematics and physics could not be separated.

As for Torricelli's own research, he, like most of Galileo's other followers, occupied himself mainly with geometry, developing Cavalieri's theory of indivisibles and solving major seventeenth-century issues such as finding the volume of infinite solids, studying the cycloid (the path followed by a point on the rim of a rolling wheel), and determining the center of gravity of any geometrical figure. In addition to his *Opera geometrica,* he wrote, and planned to publish, many other mathematical works, but died before he could do so. They were published only at the beginning of the present century.[43] The fact that these mathematical works remained unpublished for so long may have fostered Torricelli's image as an experimental physicist.

Torricelli, in fact, derived his fame from his barometric experiment, performed in 1644 with Viviani's help. By means of a column of mercury, which they varied in height to control the compression of entrapped air, they measured changes in the air: "now heavier and coarser, now lighter and more subtle."[44] One important result, creation of an artificial vacuum, was sensational. Torricelli is remembered more for this one experiment than for all his work in geometry. It also helped to consolidate the empiricist image of Galileo by virtue of his disciple's demonstration.

The experiment touched on a sensitive topic, extremely sensitive if Redondi's interpretation is correct that atomism rather than Copernicanism was the main cause of Galileo's trial.[45] Not only did the experiment refute Aristotle, it may have had far-reaching theological implications relating to transubstantiation. Torricelli, probably aware of such possible ramifications, was cautious and diplomatic; he did not publish a description of the experiment but described it in a letter to Michelangelo Ricci. The choice of Ricci was not casual; Ricci, a former pupil of Castelli and of Torricelli himself, could be trusted as both a scientist and a friend. Ricci also corresponded with many scientists in Europe and could spread the news. Last, and most important, thanks to his high rank at the papal court, Ricci knew how to deal with such a delicate matter.

The barometric experiment was actually one of the few instances in which Torricelli occupied himself with physics. To outward appearances, at least, he totally avoided astronomy. The question arises: Did he do so out of caution, or was he simply following a planned scientific program that reflected his personal interests? In my opinion, it was a mixture of both. In the next chapter I will try to show that the efforts, overall, of Galileo's collaborators focused mainly on geometry, and may well have been the same even without the limitations imposed by the Church. Nevertheless, Torricelli was undoubtedly cautious in what he did, said, and published. For example, evidence in his correspondence indicates that he did make some astronomical observations. Baldassar de Monconys of France, King Louis's counselor who traveled in many countries and visited philosophers, including Torricelli, all over Europe, recorded in his diary in 1645 that Torricelli was engaged in astronomy and cosmology. Torricelli was apparently not very eager to talk about these subjects.[46] On one occasion, he was challenged by Mersenne and Pierre de Carcavy, Louis's librarian, to state his opinion of a 1644 work by Roberval (on Aristarchus) and implicitly of the Copernican system.[47] Torricelli was evasive and first pretended he was not competent enough. When Carcavy persisted, Torricelli cut him short, saying "ad me minime attinet" ("It's none of my business").[48]

Torricelli justified his "neglect" of astronomy by pleading a lack of time for observations. True or not, Torricelli's reluctance to write on astronomy is understandable. After all, Galileo had been sentenced to life imprisonment for Copernican views, and his pupil could expect similar punishment. Torricelli's caution exemplifies the strange reality in which Italian scientists worked after Galileo's trial, resembling that of a modern totalitarian state. They were not free to do what they

Figure 3.3. Vincenzio Viviani. (Collezione Gioviana in the Museo di Storia della Scienza, Florence)

wanted but at the same time were not compelled to work underground. Formally, they were requested only to refrain from upholding the Copernican view in print, but they were also aware that it was not advisable to "irritate" pious people by airing comments on sensitive subjects such as the atomic theory of matter. Yet research in these fields appeared to hold no danger, and some Galilean scientists made astronomical observations without hindrance. For example, Vincenzio Renieri compiled tables of the motions of Jupiter's satellites.

But the Tuscan court and scientists in its service continued to be cautious. After Torricelli's death, in 1647, the grand duke did not appoint a court mathematician in his place, perhaps because he could not find one of Torricelli's stature with his sense of diplomacy. Cavalieri and Renieri also died in 1647, leaving Viviani as the foremost Tuscan scientific authority. Viviani was still relatively young (25 years old) and had not yet had an opportunity to prove himself. The chairs

Figure 3.4. Giovanni Alfonso Borelli. (Collezione Gioviana in the Museo di Storia della Scienza, Florence)

of mathematics in both Pisa and Florence remained vacant, and scientific activity at the Tuscan court centered around the interests of Ferdinand and Leopold, the Medici brothers who ruled the country.[49]

The main scientific objective of the Tuscan court in the decade following Torricelli's death was to publish Galileo's and Torricelli's works. The result was publication of a partial collection of Galileo's writings in 1655–1656 in Bologna. Various scientific experiments were occasionally performed at court, but activity was sporadic and merely a prelude to the more regular research carried out between 1657 and 1667 by Borelli and the Accademia del Cimento.

In 1656 Borelli received the chair of mathematics at the University of Pisa. He was then nearly fifty years of age and had behind him a long and successful scientific and academic career in Sicily. His arrival coincided with the beginning of the activity (detailed in chap. 8) of the Accademia del Cimento, in which Borelli was a major participant.

Thus, Galileo's major followers contributed mainly to geometry, particularly the field of indivisibles. Cavalieri and Torricelli devoted most of their life to this topic. Nardi, Ricci, Viviani, Magiotti, and Borelli also dealt with it, to a greater or lesser extent. An understanding of this work of Galileo's followers naturally requires consideration of its basis: Cavalieri's theory of indivisibles. But even this is too advanced a starting point. One must begin with Galileo's study of the continuum.

4

The Indivisibles

Galileo and the Continuum

Galileo is famous mainly for his telescopic discoveries, his campaign for Copernicanism, and his theory of motion. Less known, but perhaps no less important, was his study of the continuum, both geometrical and physical,[1] which played a significant role in his and his followers' work. For example, Galileo used infinitesimal geometry based on continuum concepts to prove the law of free fall, one of his major discoveries.[2] Nevertheless, the Galilean view of the continuum is far from clear, a condition that mirrors the difficulties always associated with this subject.[3] A brief outline of the work on the continuum done by Galileo and his followers will indicate both the importance of the subject and how attendant problems led the Galileans into a crisis. The crisis, despite contrary assertions by the modern historian Pietro Redondi, seems to have been purely conceptual and to have had little or nothing to do with religion.

The term "indivisible" (normally used in geometry in the seventeenth century) is the Latin version of the Greek word "atom" (normally used in physics and chemistry), and in Galilean thought the geometrical and the physical continua were strictly interrelated.[4] The effective mathematical technique of regarding geometrical shapes as made up of an infinite number of infinitely small (infinitesimal) parts raised the classic question of how to shift from the finite to the infinite. The analogous physical question was equally problematic: Is matter continuous and infinitely divisible, as Aristotle thought, or composed of a great but finite number of indivisible particles, as Leucippus and Democritus thought? Also, how is the geometrical continuum related to the physical continuum?

These questions must already have occupied Galileo in the early days of his scientific work; in his *De motu* he discussed the properties of various types of media in relation to motion, and even spoke of particles.[5] He had probably planned eventually to treat the subject of the continuum thoroughly when he listed it as an item in the research program he presented to the Tuscan court in 1610. Yet he never wrote, or at least never published, such a work and, amazingly, only occasionally dealt with atomism in print. These occasions often directly involved his followers.

The first instance, concerning the nature of the physical continuum, occurred in 1611 in his controversy with Lodovico delle Colombe over floating bodies. To Colombe water was a nonatomic continuum, proved by experience showing that moving one part of it also set other parts in motion. Galileo refuted him with a logical argument leading to an absurd conclusion. He said that the reasoning Colombe applied to water could also show that a heap of grains was actually nongranular (i.e., continuous).

> For although of a heap of wheat, which is an aggregate of discontinuate parts, one can move a single grain without moving others, this will not be done recklessly by throwing a stone [as in Colombe's example], or by stirring it with a rod. . . . Whoever wants to move a single grain must touch one alone with a little pick, and drive it to one side with great care—and so much the greatest delicacy is required as the little component bodies are still smaller; hence I think it is going to take Sig. Colombe a lot of trouble to separate one at a time his grains of cinnabar and lapis lazuli. So you see how vain and irrelevant is Sig. Colombe's experience to prove the continuity of water by throwing a stone into it and observing that the motion of the first parts touched by the stone then moves other parts.[6]

As on later occasions, Galileo was being cautious, expressing only an implicit favor for atomism. In a second essay on the same controversy, written with Castelli, he even argued that he had never said the continuum is composed of indivisibles.[7] From the very beginning, then, Galileo publicly displayed a certain ambiguity, or at least uncertainty, about atomism.

But a few years later, between 1616 and 1620, Galileo must have been instrumental in inspiring Cavalieri to investigate the geometrical continuum, since he was Cavalieri's reference point, and Cavalieri acknowledged his debt in books and letters. During this period, Galileo also entered another controversy, much more serious, perhaps more dangerous, and at least indirectly related to the continuum. This dispute was with Father Orazio Grassi, the Jesuit mathematician of the Roman College who had claimed that comets were real celestial bodies farther away than the moon. Galileo held that comets were only optical effects, and Mario Guiducci, the consul of the Florentine Academy, presented Galileo's views to the academy in 1619 in *Discourse on the Comet.* Guiducci, a follower of Galileo, was well qualified to represent Galileo against Grassi, having studied at the Roman College and also been a pupil of Castelli. *Discourse on the Comet* was published soon after.[8]

Grassi replied at the end of the year with his *Libra astronomica ac philosophica,* published under the pseudonym Lothario Sarsi.[9] He relied on optical arguments, among others, a field in which he was well versed. As Redondi has pointed out, the focus of the controversy shifted from astronomy to the physics of heat, light, and fluids.

Galileo replied to Grassi in his *Assayer* (1623), which was also a general outline of Galileo's scientific thought, writing explicitly that perceivable effects such as light and heat can be explained by means of particles. These remarks, as Redondi discovered, were judged heretical by a pious Roman who, a year later, sent an anonymous complaint to the Inquisition.

Redondi's findings may possibly illuminate Galileo's growing ambiguity in the following years. First, Galileo began to express misgivings about Cavalieri's theory. Second, from 1623 to 1638 (the year the *Two New Sciences* appeared) he published nothing on the subject even though his notes and letters show he was at least thinking about it.[10] Third, in 1626 Cavalieri asked Galileo repeatedly how his work on indivisibles was proceeding.[11] Was Galileo writing such a work? If so, why is there no trace of it? Did Galileo feel threatened? Galileo's specific criticisms of Cavalieri and when he made them are unknown because most of his letters to Cavalieri were lost. Yet he most probably expressed his doubts before publishing the *Assayer,* and no evidence indicates that they were based on other than purely mathematical grounds. In general, despite the attractiveness of Redondi's implying Galileo may have been intimidated by fear of arousing the Inquisition, the difficulties seem to have been purely conceptual, with no signs involving religion.

In 1633 Antonio Rocco, a renowned Aristotelian philosopher in Venice (not to be confused with Giannantonio Rocca, a mathematician and Cavalieri's friend) published a book criticizing Galileo's *Dialogue,* saying that any continuum, in philosophy or geometry, is infinitely divisible.[12] Galileo took issue with Rocco's views in some notes he wrote in the book's margins.[13] The notes were a starting point for his treatment of the continuum in his *Two New Sciences,* which clearly disclosed Galileo's conceptual difficulties.

In the *Two New Sciences* the "infinitesimal" mathematics (based on indivisibles) was part and parcel of physics. The first two "Days" of the book were devoted to the fracture resistance and causes of cohesion of solid bodies and treated the bodies as if they were a mathematical continuum. The Third and Fourth Days discussed motion, and Galileo used the theory of indivisibles to prove the law of free fall.[14]

The First Day also presented a general treatment of infinitesimals,[15] beginning with a discussion of the relation between a line and the points that compose it:

> If we consider the line resolved into an infinite number of infinitely small and *indivisible* parts, we shall be able to conceive the line extended indefinitely by the interposition, not of a finite, but of an infinite number of infinitely small indivisible empty spaces.[16]

Then he considered the two- and three-dimensional cases:

> Now this which has been said concerning simple lines must be understood to hold also in the case of surfaces and solid bodies, it being assumed that they are made up of an infinite, not a finite, number of atoms. Such a body once divided into a finite number of parts it is impossible to reassemble them so as to occupy more space than before unless we interpose a finite number of empty spaces, that is to say, spaces free from the substance of which the solid is made.[17]

By introducing "empty spaces" Galileo began to shift from the geometrical to the physical problem, and to enforce the analogy by considering a small ball of gold. But the reasoning quickly evoked questioning by Simplicius (the book's Aristotelian interlocutor):

> This building up of lines out of points, divisibles out of indivisibles, and finites out of infinites, offers me an obstacle difficult to avoid; and the necessity of introducing a vacuum, so conclusively refuted by Aristotle, presents the same difficulty.[18]

Galileo soon admitted that indivisibles "transcend our finite understanding," and argued in a long and captivating discussion that they nevertheless can be used to calculate areas. But he concluded with the admonition:

> These are some of the marvels which our imagination cannot grasp and which should warn us against the serious error of those who attempt to discuss the infinite by assigning to it the same properties which we employ for the finite, the nature of the two having nothing in common.[19]

Thus, until his last published work he remained inexplicit on the question of the continuum.[20]

Galileo's ambivalence was noted by his followers, for example, Nardi in his *Scene:*

> Now Galilei, although speaking very ambiguously of the actual and potential infinities, of number and weight, as also of the continuum and of assemblage [*congiunto*], and of other similar principles, he nevertheless lets understand that bodies are composed of actual indivisibles . . . but this principle, with other consequences, should be clarified.[21]

Can the work of Galileo's followers tell us something more than that of Galileo alone? The study of the continuum, in both geometry and physics, was a substantial issue for some of them. In mathematics, Cavalieri devoted his life to developing the theory of indivisibles, Torricelli applied it, and Nardi paid much attention to it. In physics, Torricelli's results with vacuum experiments implied the existence of atoms, and the unpublished papers of the Accademia del Cimento show that atomism was a major topic in its research. Interestingly, Cavalieri and Torricelli did not consider geometrical indivisibles and physical atoms simultaneously as Galileo did, but this seems to reflect neither a different approach nor an attempt to avoid possible theological objections. Rather, their interest lay primarily in geometry, and their efforts in physics were relatively marginal. Here we will first concentrate on Cavalieri's theory of geometrical indivisibles and its application by Torricelli, leaving the physical continuum for later chapters.

The study of the geometrical continuum was one of the main pursuits of Galileo's followers, and their efforts can be said to have been both a success and a failure. As a tool to determine the volumes of variously shaped objects and as an aid to learning and discovery, indivisibles were very successful, but the Galileans failed to dispel serious doubts about the nature of the indivisibles and the foundation of the whole theory. Perhaps the task of elucidating the theory was too great, since the foundations of Newton's and Leibniz's infinitesimal calculus later encountered similar criticism, most notably in 1734 by the Irish bishop and philosopher George Berkeley. (After many attempts to make calculus more rigorous, it is still not certain today that the basic problems have been solved.)

Cavalieri's and Torricelli's Indivisibles

Cavalieri's mathematical theory of indivisibles, expounded in his *Geometria* (1635) and *Exercitationes* (1647), had important practical and heuristic advantages but, at the same time, fundamental theoretical disadvantages. Its most important advantage was to facilitate the calculation of areas and volumes and hence the solution of some major geometrical problems. Its main disadvantage was its reliance on concepts such as indivisibles whose definitions did not satisfy most contemporary mathematicians, even those who developed and used it.

At the beginning of the *Geometria,* Cavalieri said he was attempting to expand the work of Greek forerunners, to measure the area and volume of as many plane and solid geometrical figures as possible. The project was more ambitious than it might seem. The classical rigorous approach was the method of exhaustion, "measuring" an unknown area or volume by showing it was neither larger nor smaller than one that was known, an a posteriori and laborious procedure. In Cavalieri's time the measurement of surfaces and volumes was still the major problem of geometry, and he believed that the Greeks had less cumbersome techniques.[22] Cavalieri was actually trying to reinvent such a technique.

The *Geometria* was divided into seven books, the second, introducing the concept of indivisibles, being the most important. Cavalieri's approach, called "semi-atomistic" by Lucio Lombardo-Radice, the modern translator-editor of the *Geometria,* was essentially intuitive and would hardly satisfy a modern mathematician.[23] Cavalieri let a straight line move, without changing its angular orientation, over a surface, and the "trace" the line left, called "all the lines," was the infinity of "indivisibles" that compose the surface. Since all the component lines were parallel to each other, their common direction could be represented by any one of them, which Cavalieri called the *regula.* (In three dimensions, an elemental surface moved in one direction to trace out a volume.) His basic proposition, expressed schematically, was that if A and B are the areas of two surfaces A and B, and L(A) and L(B) are, respectively, *all the lines* in A and B, then:

$$A/B = L(\mathrm{A})/L(\mathrm{B}).$$

But Cavalieri did not explain how to "measure" an aggregate of infinite numbers of lines or the difference between a simple area and "all the lines." A modern reader may be tempted to consider the concept of "all the lines" as an infinite set in group theory, or an integral in calculus. However, group theory and calculus were developed much later.

Despite this lack of clarity Cavalieri achieved some impressive results. The best known, called Cavalieri's theorem, originally dealt with a specific case but can be generalized in modern calculus notation (a formulation Cavalieri would not have recognized at all) as:

$$\int_0^a x^n \, dx = a^{n+1}/(n + 1)$$

where $n = 1, 2, 3, 4, 5, 6, 9$.[24] He also succeeded in calculating with ease the volumes of a large number of solids of different shapes and pointed out the applicability of his theory to other fields, such as physics.

The merits of Cavalieri's theory were soon recognized, and Nardi noted in his *Scene* that "the very ingenious principle of the indivisible nowadays emerges with great success."[25] Its main success came in its mathematical applications by Torricelli.

Torricelli extended Cavalieri's theory in various ways. In particular, he developed the use of "curved indivisibles," a type Cavalieri had already introduced in the *Geometria* but without a precise definition. Cavalieri admitted that this was a matter in which Torricelli had surpassed him.[26] Torricelli achieved a series of impressive results, which he included in his *Opera geometrica* in 1644. For example, he showed that the area of a hyperbolic solid, which extends to infinity, is actually finite; he found various ways to square the parabola; most important, he proved that the area under a cycloid—the path described by a point on the circumference of a circle rolling on a straight line—and above the line equals three times the area of the circle.[27] In 1646, in a letter to Michelangelo Ricci, Torricelli also presented a general principle for finding the center of gravity of plane and solid figures having an axis of symmetry.[28]

Although far from complete, this account of the main achievements of Cavalieri's and Torricelli's theory of indivisibles shows that it is fair to conclude that it fulfilled the objective announced at the beginning of the *Geometria*. But achievements in applying the theory were only partly matched by Cavalieri's rational systematization of the concept of indivisibles. He advanced somewhat beyond earlier primitive explications, but basic questions that had also troubled Galileo remained unresolved and were soon to cause strong opposition.

The Crisis

Cavalieri did not, or could not, clearly state how his infinitesimal mathematics could compare infinite magnitudes (the numbers, so to speak, of infinitesimal lines that generated surfaces, or of infinitesimal plane surfaces that generated solids) with finite magnitudes (the ordinary areas and volumes he was calculating). It was this weakness that Paul Guldin (1577–1647) vigorously attacked soon after the theory was published.[29]

Guldin, a Jesuit mathematician who had studied and taught at the Roman College, was a friend of Orazio Grassi (with whom Galileo had had a dispute) and, like Grassi, had openly supported Galileo in 1616 when Galileo was instructed to abandon the Copernican theory.[30]

Also like Grassi, Guldin was involved in a number of controversies. Guldin's main work, *Centrobaryca,* dealt with, among other subjects, the determination of the center of gravity of plane figures and contained the theorem (that carries his name) on the volume of a solid generated by a plane figure revolving about an external axis. *Centrobaryca* was controversial. In it, Guldin criticized Kepler's book *Stereometria doliorum* on infinitesimals (1615), claimed that Cavalieri had plagiarized from both Kepler and the mathematician Bartholomew Sover, and pinpointed the major weakness of Cavalieri's theory in its basic concept of "all the lines," saying that this infinite quantity cannot be compared with a finite measurement.

Cavalieri in his *Exercitationes,* written in response to Guldin's criticisms, more or less satisfactorily countered the accusations of plagiarism but did not dispel the doubts about his theory's foundation. All he was able to maintain was the existence of a type of infinites—including "all the lines"—that are subject to some kind of finiteness and that the two can be compared with each other, a lame argument that could hardly refute Guldin's objections.

Guldin's criticism drove Galileo's followers into an intellectual crisis. In their correspondence they continually referred to it but could not devise a decisive answer. In 1643 Cavalieri asked Torricelli to send him some of Torricelli's results with curved indivisibles, hoping to publish them with his own work on trigonometry and logarithms as an argument against Guldin, but Torricelli declined to cooperate.[31] Whenever he could, Torricelli avoided the use of indivisibles, for example, in his treatise ("De sphaera et solidis sphaeralibus") on spherical solids, included in the first part of his *Opera geometrica.* Cavalieri, astonished, expressed disappointment: "I confess that seeing you avoiding [indivisibles] totally in your work concerning spherical solids raised in me some fear to have lost such a celebrated supporter."[32]

Cavalieri, Torricelli, Nardi, and other Galilean supporters of the theory of indivisibles could only acknowledge its weakness. Torricelli did it in the most open way: Not long before his death he wrote two short treatises entitled "Contro gli infiniti" ("Against the Infinities," citing instances of the incorrect use of indivisibles) and "De indivisibilium doctrina perperam usurpata" ("The Doctrine of Indivisibles Used Improperly"), which were not published until the early years of the twentieth century.[33] On another occasion he remarked "That the indivisibles are all equal, i.e., point to point, line in breadth to line and surface in depth to surface, is an opinion which I find not only to be hard to prove, but even false."[34]

Nardi was even more perplexed. At the beginning of his *Scene* he wrote:

> The universal and intimate subject of mathematics (namely the quantum with its modes and cases) falls within metaphysical contemplation and its precise essence cannot be grasped as it is by the human mind. Hence its definition and the ultimate cause of the continuum remain unknown. This is because we have supposed that neither can the senses reach the surface of things, nor can intuition be reduced by our mind.[35]

This passage not only declared Nardi's feeling of helplessness in defining infinitesimal quantities, but also described indivisibles as neither physical nor geometrical but rather metaphysical entities, and showed that Galileo's followers did not consider them purely geometrical.

Later in the work Nardi added:

> In the wake of the mathematicians, I assumed that in theory the lines, the points, and the surfaces exist in bodies as true parts and components, hence these terms really exist and can, by means of some force, be separated from the body, and by means of some supreme force, all be separated. Hence there would be a space and an infinite number, which is impossible.[36]

(It is interesting to note that Nardi here also speaks of points generating lines, in addition to the previously discussed lines generating surfaces and surfaces generating solids.)

In other words, according to Nardi, the geometrical concept of indivisibles cannot be transferred to physics, as Galileo, Cavalieri, and Torricelli may have wanted, and cannot be used to measure physical surfaces. Nardi, then, implicitly admitted that Guldin was right.

Other scandals may also have added to the dismay of Galileo's followers about the theory of indivisibles. In 1646, the French mathematician Gilles Personier de Roberval claimed priority as to Cavalieri's theory of indivisibles and Torricelli's theory of the cycloid. This initiated a long dispute, continuing after Torricelli's death, that involved mathematicians all over Europe—Galileo's Tuscan followers, Roberval's French supporters, including Blaise Pascal, and John Wallis in England—and became a matter of national pride.[37]

One can well see how onerous were the open questions Galileo left his followers, and why they were so perplexed and often frustrated. The difficulty with the theory of indivisibles indeed seemed to be, above all, conceptual, and Galileo's followers may have been able to defend formal priorities but not to reject the substantial, everlasting criticism. There is no sign, however, that religion had anything to do with all this—as suggested by Redondi—although clearly atomism may have held more importance for the Galilean school than is commonly thought.

Traditional literature regards the religious factor as the main cause of the decline of Italian science during the second half of the seventeenth century. It disregards conceptual factors, such as the dominant arguments over the nature of the geometrical continuum, which certainly encumbered scientific progress. And these were not the only "internal" difficulties, the next chapter describes others, having more to do with physics and methodology.

5

Torricelli's Rationale

We now return to methodology. In discussing the essence and structure of science, Alexandre Koyré argued that conceptual, as opposed to purely empirical issues predominate in modern science or, at any rate, that this is how a historian of science should see it. According to Koyré, modern science, unlike natural philosophy, consists of theories that precede and determine experiments, which, in turn, confirm the theories. Thus, while crude observation is typical of natural philosophy and played only a minor role (may even have been an obstacle) in the evolution of modern science, experimental activity based on preconceptions nourished the growth of science. Hence modern science, according to Koyré, is *experimental,* rather than *experiential.*[1]

Koyré's distinction between inductive, experiential natural philosophy and a priori science perhaps lacks analytical clarity and may be somewhat misleading. But a similar distinction was already made in Galileo's day. Cavalieri, for instance, was well aware of the possibility of investigating nature in the two different ways and, in the introduction to his *Lo specchio ustorio* (1632), clearly distinguished between an inductive approach:

> The way Mercury learnt from the tortoise how to make the lyre, or Pythagoras invented music from striking hammers, i.e., in a way to know the conclusion in advance and then to investigate from it the principles;[2]

and an a priori, hypothetical-deductive approach, through which

> Columbus thought that for this and that reason the new Indies existed, and their discovery was going from principle to conclusion. And this is what occurs here.[3]

Cavalieri even apologized for presenting an "a priori" work, indicating that contemporary intellectuals preferred inductive empirical sciences. "He who knows my many occupations will forgive me and accept for the time being the speculative part, and later will see how easily it works in practice."[4]

Although Cavalieri's readers may have favored an inductive approach, modern science, according to Koyré, remains a priori, just as Cavalieri presented it in his *Lo specchio ustorio*. To confirm this judgment Koyré studied in depth the works of some of the leading figures of the scientific revolution, such as Galileo, Kepler, Descartes, Borelli, and Newton,[5] touching only marginally on Cavalieri and Torricelli, although Torricelli in particular might have provided additional interesting cases worthy of study.

I will argue that although Torricelli may not have been an a priorist, in a nontrivial sense of the term, he was certainly not a mindless or uncritical empiricist either. His methods tended to be hypothetical-deductive. My grounds for this belief are: (1) He designed his barometer experiment to test and choose among three available theories; (2) His response to empirical criticism of Galileo's theory of projectiles was not to abandon the theory but rather to represent it as a piece of pure mathematics; and (3) He responded similarly to a specific criticism of the projectile theory by an artillery man (G. B. Renieri) and also suggested further experiments. First we will examine his best-known project: the barometer experiment.

The Barometer

The description of the barometer experiment appeared for the first time in Torricelli's letter of June 11, 1644 to Michelangelo Ricci, which was published in 1663 by Carlo Dati in his *Letter to Filaleti*. It has since become well known throughout the world, translated into many languages.

Much has been written on the experiment, so an exhaustive treatment is not warranted here.[6] The present purpose is primarily to determine what this famous experiment can reveal about Torricelli's and, possibly, Galileo's approach to science.

In his early manuscript *De motu* Galileo argued, against Aristotle, that motion can take place in a void (vacuum).[7] Thus, from the beginning of his scientific career, he took it for granted that a vacuum can exist and was concerned only with its properties. He later considered the relation between a vacuum and the pumping of water and concluded (wrongly) that water was pumped only by the action of a vacuum.

Yet, as early as 1630 Giambattista Baliani, a Genoese official with wide scientific interests, suggested the correct explanation. In a lucid letter to Galileo, Baliani agreed that a vacuum could be produced (his expression "si desse" meant "would let itself") but only with difficulty, and he therefore believed it was the atmospheric pressure that raised the water. Baliani, however, admitted not being able to visualize this experimentally. In his letter he also said: "We are under the immensity [of air] and do not feel its weight and pressure from all sides."[8] Torricelli would later express himself in similar terms and may well have taken his cue from Baliani.

Galileo, nevertheless, insisted that a vacuum had a "force" [*forza del vacuo*] or a "resistance," and that it was the vacuum in a water pump that exerted a "resistance" on water equivalent to the weight of a column of eighteen ells (about ten meters).[9] He also said, "Whenever a cylinder of water is subjected to a pull and offers resistance to the separation of its part this can be attributed to *no other cause* than the resistance of the vacuum."[10]

Did Galileo reject Baliani's explanation (later proven correct by Torricelli) only because it "diminished" the importance of vacuum on which Galileo so insisted? It is difficult to say, but such a motive would not have been exceptional for Galileo.

There were thus, within the Galilean circle, at least two different explanations for raising water: Galileo's and Baliani's. And there was, of course, the traditional Aristotelian view, opposed to both and held by the Jesuit scientists, among others, that air has no weight and a vacuum does not exist. For Aristotelian philosophers the pumping of water was relatively easy to "explain," as the consequence of the removal of matter from the space above the water: Since there could be no vacuum, water would immediately replace the removed matter. The appearance of Galileo's *Two New Sciences* may have stimulated all parties to further investigate the subject. An experiment could, perhaps, decide which explanation was more correct. (Both Baliani and Galileo cited numerical results, indicating reference to some form of empirical test.)

Evidence indeed exists of experiments on water and vacuum performed in Rome between 1639 and 1643, probably in 1641, when Galileo was still alive. Rome at that time was still the main center of the Galilean school, and Castelli, its founder, was also the leading scientist in matters of hydrostatics. Castelli may have encouraged these experiments, although I have found no evidence of this. The experimenters were nevertheless Roman followers of Galileo: Gasparo Berti, with the help, or at least in the presence, of Raffaello Magiotti and two distinguished Jesuit professors, Athanasius Kircher and Niccolò Zucchi.

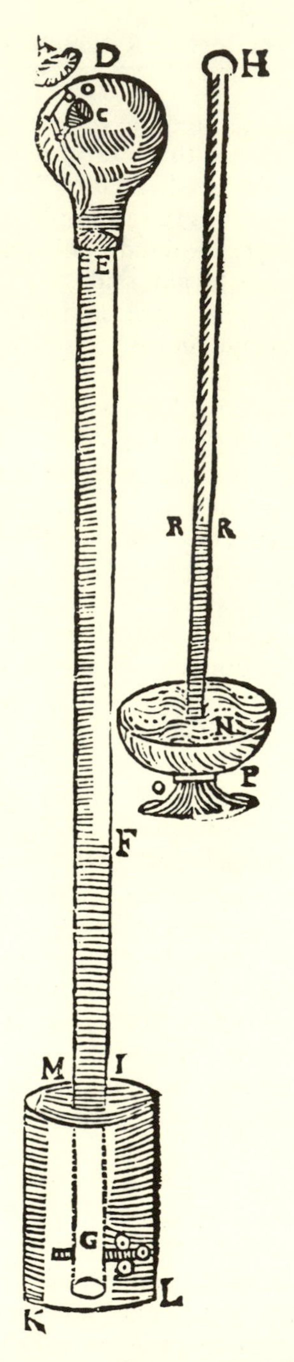

Figure 5.1. Vacuum apparatus: Gasparo Berti (left) and Evangelista Torricelli (right). (Athanasius Kircher, *Musurgia universalis*) According to Kircher, Berti also used a bell (OC) to test the vacuum.

Several descriptions of these experiments are known, all written several years later, after Torricelli had performed his barometer experiment and found the correct explanation.[11] They depict a siphon in which a column of water (not mercury) flows from a closed tap through a tube down into a cask. The reports were similar to Torricelli's and, since they were written after his, may well have been anachronistic. Also, they did not clearly indicate the exact purpose of Berti's experiments. The historian Mario Gliozzi, in a detailed article on Torricelli's experiment in the last volume of Torricelli's collected work (1944), once even proposed that Berti performed his experiment after Torricelli's.[12] However, Berti died in 1643, and unless Torricelli did his experiment earlier—which is unlikely—Berti's came first.[13]

Significantly perhaps, Torricelli had been a member of the scientific circle of Castelli, Berti, and Magiotti; C. de Waard, a twentieth-century historian, even discovered in Vienna a letter from Magiotti to Mersenne saying he had written to Torricelli about Berti's experiment.[14] Magiotti wrote that Berti was trying to convince Galileo, implying that Berti disagreed with Galileo and may even have sided with Baliani. If Magiotti had to write to Torricelli, and if Galileo was still alive at the time, then Torricelli had probably already left Rome to assist Galileo in Tuscany, in which case the date of Berti's experiment was probably near the end of 1641 (Torricelli worked with Galileo in the last three months of 1641).

What happened between 1641 and 1644? Were there more experiments in Rome or in Florence? Torricelli's experiment could well have been part of a joint effort of Galileo's followers to study the relation between vacuum and the raising of liquids. Torricelli may also have seen Baliani's letter to Galileo or had other contacts with the Roman Galilean school. If so, in 1644 he had one more reason to announce his success (in his June 11 letter) to none other than Galileo's leading Roman follower, Michelangelo Ricci (Castelli, like Berti, had died a year before). Evidence is lacking for any such contacts, or even that Torricelli ever received the letter Magiotti told Mersenne he had written. All the available documents show is that suddenly, on June 11, 1644, Torricelli sent the description of his experiment to Rome. He and his collaborators—Viviani, at least, was one of them—may well have reached their result independently.

In his June letter, Torricelli reported to Ricci that "some sort of philosophical experiment was being done concerning vacuum; not simply to produce a vacuum, but to make an instrument which might show the changes of the air, now heavier and coarser, now lighter and more subtle."[15] He cited several views then current concerning vacuums—the Aristotelians': "Many have said [that vacuum] cannot happen"; Baliani's (though without naming him): "others that it happens, but with the repugnance of nature, and with difficulty"; and added: "I really do not remember that anyone has said that it may occur with no difficulty, and with no resistance from nature," apparently referring to Galileo's own belief that a vacuum is not easy to create. He then rejected Galileo's view, saying "It would seem vain to try to attribute that resistance to the vacuum itself," and also commented disparagingly on "some philosopher [unfortunately unnamed], seeing that he could not escape confessing that the gravity of air is the cause of the resistance that is felt in producing vacuum, would not say that he conceded the operation of the weight of the air, but would persist in his assertion that Nature also helps by her repugnance to the vacuum."

Finally, echoing Baliani, Torricelli said: "We live submerged at the bottom of an ocean of elementary air which is known by incontestable experiments to have weight, and so much weight that the heaviest part near the surface of the earth weighs about one four-hundredth as much as water."

Torricelli's remarks pointedly reveal that contemporary opinions about the existence and behavior of a vacuum were many and various.

Torricelli then described his experiment: A two-ell (about 120 cm)-long glass vessel (or tube) sealed at one end was filled with mercury and, its mouth stopped with a finger, was turned upside down (closed end up) in a bowl of mercury. The mercury in the tube descended to a height of about one ell, leaving an empty space at the top. To prove that the space above the mercury was indeed empty—perhaps the most important part of the experiment—Torricelli added water to the mercury in the bowl and then raised the tube slowly. When the mouth of the tube rose to the surface of the water, the mercury in the column poured down, and the water rushed in to fill the tube to its top.

The fact that water rushed up to fill the tube was for Torricelli satisfactory proof that the space had been empty. However, it may not have convinced the Aristotelians: In 1650 Athanasius Kircher, basing his judgment on Berti's experiment, still claimed that the Torricellian space (see Fig. 5.1) was not empty because sound could be transmitted in it.[16] Kircher argued (rightly) that sound could be transmitted only in matter. Since the "empty" space was only a partial not an absolute vacuum, Kircher's interpretation of Berti's experiment was correct, indicating how equivocal an experiment can be.[17] These substantive doubts of the Aristotelians may have induced Torricelli and Viviani to perform additional experiments. Dati, in his *Letter to Filaleti,* reported attempts to introduce small animals, fishes, large flies, and butterflies into the top part of the tube to see if they could survive in the "empty" space, but the creatures were entrapped in the mercury, thus thwarting the test.[18] Torricelli, however, did not change his mind.

Having succeeded in producing a vacuum and refuting Aristotle, Torricelli's next step was refuting Galileo. The question was: What sustained the mercury in the column at a height of one ell above the surface of the mercury in the bowl? According to Galileo, it was the "force" of the vacuum. According to Baliani and Torricelli, it was the pressure of atmospheric air on the surface of the mercury in the bowl.

Torricelli had to test two hypotheses. To decide which was correct he performed a "crucial" experiment, repeating his previous procedure, but with two tubes (A and B in fig. 5.2), one ending in a large bulb at the top. The mercury dropped to the same level in both tubes. Had the vacuum exerted the force, as Galileo thought, "The vessel AE would have had more force, there being more rarefied attracting stuff, and this much more vigorous by virtue of its greater rarefaction than that in the very small space B." The result of the second experiment disproved Galileo's theory and favored Torricelli's and Baliani's.

Having demonstrated the effect of atmospheric pressure, Torricelli applied it to the hypothetical case of a column of water. "Water," he said, "in a similar vessel but very much longer, will rise to about eighteen ells, that is to say, as much higher than the quicksilver as the quicksilver is heavier than water, in order to come into equilibrium with the same cause, which pushes the one and the other." Torricelli thus not only explained what Galileo had observed and described, but also conjectured the hydrostatic principle, later elaborated by Pascal, according to which the ratio of the heights of two nonmixing liquids in balance is inverse to the ratio of their densities.

Torricelli ended his letter by admitting that he had not succeeded in his chief intention "to find out with the instrument EC when the air is coarser and heavier and when more subtle and light; because the level AB changes from another cause (which I never thought of), that is, it is very sensitive to heat and cold, exactly as if the vase AE were full of air." This last sentence may sound a bit obscure; Torricelli wanted to say only that his apparatus was too sensitive to heat and cold to be used as an accurate instrument to measure variations of air pressure.

What can we now judge to be the essential importance of Torricelli's experiment?

Historians emphasize that the experiment refuted the Aristotelian belief that a vacuum cannot exist. Hence, the invention of the barometer may be considered as important as the invention of the telescope

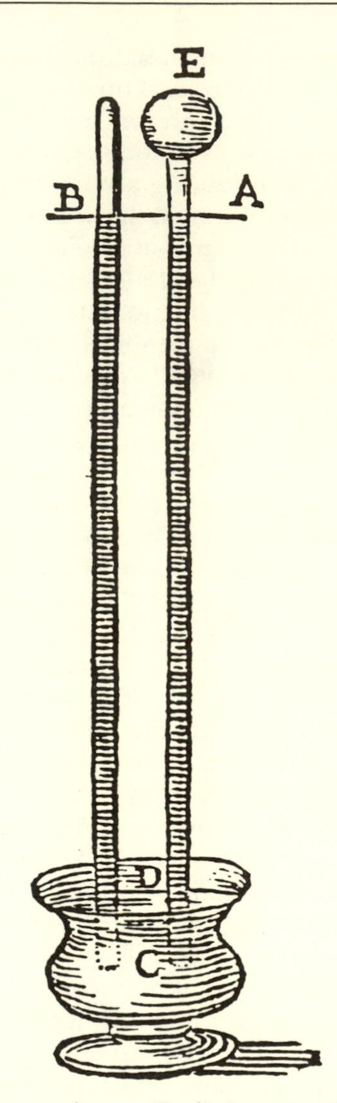

Figure 5.2. Torricelli's experiment. (Dati's *Lettera a Filaleti*)

or the microscope. This opinion is by no means exaggerated, and Torricelli was well aware of it. It seems clear, however, that Torricelli, like Galileo before him, took for granted that a vacuum could be produced—he said explicitly that he intended to do more than create a vacuum—and was perhaps more intent on *challenging* Galileo's assertion that a vacuum exerts "force" than Aristotle's that it cannot exist.

Torricelli therefore designed the barometer experiment to test previous theories rather than to generate new ones. It was a typical case of what Koyré calls experimental, as opposed to experiential, empiricism, and the philosopher of science Carl Hempel cites it as a classical example of experimentation as a method of testing rather than discovery.[19] Dati even said in the seventeenth century that "Torricelli did not come across his experiment by chance, but was guided by a clear thought, and by the time he saw and experimented the effect, he had already speculated the cause."[20]

Torricelli's experiment was, nevertheless, more than a classical crucial contest between two ideas. It was positioned at the crossroad of at least three theories, Aristotle's, Galileo's, and his own (or Baliani's), and reflects Torricelli's talent as a physicist.

At the same time, Torricelli expressed himself in terms so similar to Baliani's, and his apparatus was so similar to Berti's, that his experiment can perhaps justifiably be seen as the climax of twelve years of teamwork by an assortment of Galileo's followers: Baliani in Genoa, Berti, Magiotti, and Ricci in Rome, and Torricelli and Viviani in Florence. Torricelli may have taken Berti's device as a starting point, improved it by using mercury instead of water, and creating a vacuum to refute Galileo. Even if it were teamwork, Torricelli's contribution—conceiving a simple experiment, with simple apparatus, to test three theories conclusively and demonstrate the effects of atmospheric pressure—remains undiminished.

Both the existence of some kind of teamwork and the hypothetical deductive nature of Torricelli's methods receive confirmation in the exchange of letters between Torricelli and Ricci that followed the experiments. On June 18, 1644 Ricci replied to Torricelli's letter with a number of questions and objections (mail service was still excellent between Florence and Rome).[21] On June 28 Torricelli immediately responded with convincing counterarguments.[22]

Ricci first proposed a further test of Torricelli's results. Assuming, as Torricelli did, that air and not vacuum exerted force on the mercury, then sealing the bowl against the air should cause the mercury column to fall, Ricci reasoned. If it did not, the effect could no longer be attributed to the weight of air.

Torricelli answered that the air sealed in the bowl would retain the same density and exert the same pressure as external air and so produce the same result. Torricelli illustrated his argument by a thought experiment in which he imagined that even if a bowl full of air, or of some other compressible material such as wool, were "cut" with a knife, its internal pressure would not vary.

Ricci then asked how the air, which certainly exerts a force downward, can also exert a force upward? To this Torricelli replied that fluids, despite gravity, exert pressure in all directions.

Finally, Ricci pointed out that according to Archimedes's principle a body immersed in water does not push all the water above it but only the water it displaces. If mercury, too, pushed only an amount of air equal to its volume, how could the column of air push the mercury up? This was a rather confused argument, and Torricelli replied that his experiment represented a different case; the mercury in the tube was immersed neither in water, air, glass, nor vacuum. One can only admire Torricelli's clarity of thinking.

Does this account of Torricelli tell us anything new about Galileo's method of scientific investigation? Admittedly, Torricelli here was hardly representing Galileo, in fact was contradicting him, but this does not necessarily mean his methodology was not Galilean. Was Torricelli a "genuine" Galilean in the sense that his approach to science was the same as Galileo's? Before answering this question, it is useful to consider other occasions that constrained Torricelli to act more as a genuine "Galilean." These were occasions when he was called on to defend Galileo's ideas, most notably Galileo's theory of projectiles, which Torricelli had discussed in his *Opera geometrica.*

The Theory of Projectiles and Opposing Views

The theory of projectiles and its applications to ballistics is of crucial importance in the history of science. It was a classic example of a scientific development that straddled both mathematics and physics and also both natural philosophy and Renaissance technology, the two main cultural traditions that underlay the Scientific Revolution. Moreover, in the particular case of Galileo, Torricelli, Viviani, and other Galilean followers in the service of princes, it was an example of both the kind of theoretical and mathematical knowledge they were expected to teach and the practical knowledge of direct benefit to their employers. Many of the leading Renaissance scientists, such as Leonardo, Tartaglia, Cardano, Benedetti, Galileo, Cavalieri, and Torricelli, dealt with ballistics.

The pioneer of mathematical ballistics was Niccolò Tartaglia. In his *Nova scientia* (1537) and *Quesiti et inventioni diverse* (1546, second edition published in 1554) he discussed the dependence of the trajectory on the angle of elevation of the gun, finding a maximum range at an angle of 45 degrees—in Tartaglia's terms: 6 *ponti,* or *punti* (points), or

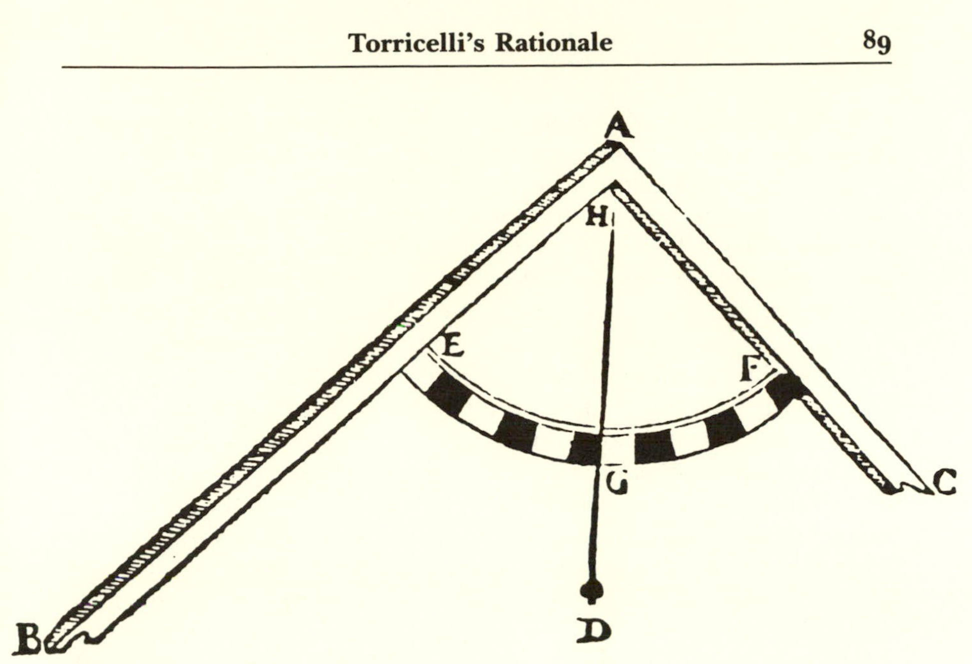

Figure 5.3. Niccolò Tartaglia's gauge.

72 *minuti* (minutes).[23] Any smaller or larger angle resulted in a shorter range. Tartaglia admitted frankly that he had never experimented with guns, which Rupert Hall believes was a factor in Tartaglia's success.[24]

Tartaglia also invented an elevation gauge (fig. 5.3) for gunnery, called the *squadra* (square, or sector). It consisted of two legs unequal in length at right angles to each other, a plumb line, and a quadrant arc (quarter of a circle) divided into twelve points, or segments; the longer leg was inserted into the gun so that the plumb line marked the angle of elevation on the graduated curved scale.

Tartaglia's writings were widely known when Galileo began his career, which included investigations in military engineering, mainly while he was in Padua. The second volume of the National Edition of Galileo's works contains two undated and unpublished treatises, "Breve instruzione all'architettura militare" (Brief Instruction in Military Architecture) and "Trattato di fortificazione" (Treatise on Fortification). The first probably dates to as early as 1592, and Favaro conjectured that the treatises were public or private courses given to Paduan students.[25] It was also in Padua that Galileo designed his geometrical and military compass in 1597, which could be used to measure, among other things, the elevation of guns. Stillman Drake has shown that Galileo derived his compass partly from Tartaglia's gauge.[26]

Galileo gave instructions on how to use the instrument in *Le operazioni del compasso geometrico et militare* (1606), his first published work.[27]

Later, Galileo developed his theory of the motion of projectiles as part of a general theory of motion. He must have outlined the theory in detail to his followers since Cavalieri included it without Galileo's permission in his *Lo specchio ustorio,* in 1632, as an example of the application of the theory of conics.[28] The motion of projectiles, Cavalieri explained, is a combination of their fall, which is proportional to the square of the time of flight, and their inertial motion, which is proportional to the first power of the time; hence its trajectory is parabolic.[29]

However, Cavalieri also remarked that a projectile's trajectory is only approximately parabolic; if it could continue to travel under the surface after impact, it would go through the center of the earth.[30] He was thinking ahead of his time.[31] If one assumes—as Newton's theory of gravitation would later postulate—that the force pulling the projectile down (the falling component) points toward the center of the earth, then (neglecting the inertial component, which could, however, be dissipated by resistance) its direction will vary as the projectile advances, and the trajectory will no longer be perfectly parabolic. But Galileo also answered Cavalieri's reservations in his *Two New Sciences.* A statement at the beginning of the Fourth Day explained that Galileo was speaking *ex suppositione,* that is, theoretically, in terms of Euclidian geometry thus rendering the motion perfectly parabolic.[32] Galileo's theory was nevertheless not entirely hypothetical. He distinguished between fast-moving projectiles, fired with the aid of gunpowder, and slower-moving projectiles, expelled from catapults, crossbows, or, at most, mortars. He devised a table to help gunners calculate trajectories, but cautioned that his theory was applicable with a relatively small margin of error only to slower-moving projectiles.[33]

Galileo's presentation seems convincing, at least to a modern reader. But it was far from convincing in his own day, since most of the other leading contemporary scientists, including his own followers such as Viviani and Nardi, criticized it.

Viviani probably expressed his objections in 1639, when he was Galileo's amanuensis, and Galileo himself mentioned them in a letter to Castelli. I could find no explicit account of Viviani's queries, but Galileo admitted having considerable trouble answering them.[34]

Nardi's criticism was thoroughgoing. As noted in chapter 2, he doubted Galileo's law of free fall, and, implicitly, his theory of projectiles. He claimed that the trajectory of projectiles is

> a mixture of two motions which do not remain pure but alternate. The violent motion appears necessarily to be faster at the exit of the piece

> than farther away, as happens in natural motion, not going through equal spaces in equal times (and here I doubt what Galileo says), and more, some mechanicians convinced themselves that the artillery ball goes for a certain space in a straight line; this is false (since the effect of gravity would be annulled), yet it seems true that at first horizontal motion sustains the projectile so that it does not fall only by the law of gravity, but later gravity must overcome the external impetus and the projectile is brought back toward the center and thus the force that had previously moved it from rest seems unable to move it in the same way after it has acquired an increased motion toward the center. Now, the combination of these motions, moments and times is a line very close to a parabola, but I find it hard to demonstrate that it is so.[35]

Nardi's analysis appears still to be influenced by Aristotelian thinking and was based on the notion of *impetus* that wears out—an indication of how difficult it was even for a follower of Galileo to discard such concepts. Yet, Nardi's was typical of the general criticisms of Galileo.

Among Galileo's leading followers only Torricelli seems to have been in general agreement with the master on these points and may therefore be taken as his best representative.[36] Torricelli completed this phase of Galileo's work in the second part of his *Opera geometrica,* entitled "De motu."[37] There he treated at length (in Latin) the motion of projectiles as an instance of Galileo's theory of motion, showed that the range increases in proportion to sin 2θ, where θ is the angle of elevation, and described (in Italian to be intelligible to gunners) a new instrument (fig. 5.4) that correlated the range and angle of elevation.[38] By doing so, Torricelli became the focus of the criticism of Galilean science.

Admittedly, Torricelli warned his readers that guns do not obey Galileo's theory, but not because the theory was incorrect, only that shooting with guns depends on additional factors, such as the type of gun, gunpowder, and cannon ball. Without these factors, he said Galileo's theory would work. Thus Torricelli hoped to convince if not gunners, at least mathematicians.[39]

But Torricelli had little success even with mathematicians and philosophers. Fundamental doubts about Galileo's and Torricelli's science were raised in France at the beginning of 1645 by Mersenne, Descartes, and Roberval. Mersenne, who corresponded regularly with Torricelli, first questioned the proposed relationship between motion and its causes, for example, for the progression of a body down an inclined plane and the flight of an arrow. In modern terms, he was asking for a dynamical explanation of kinematical questions.

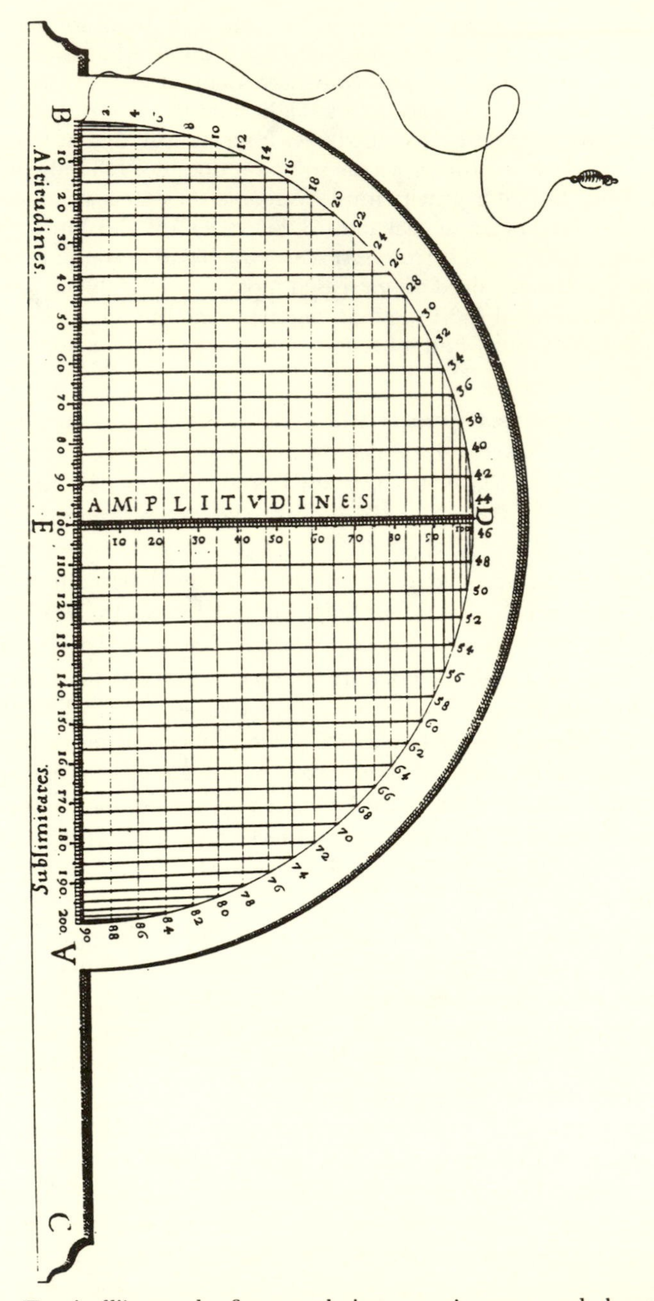

Figure 5.4. Torricelli's *squadra* for correlating a gun's range and elevation.

Torricelli was unable to go beyond kinematics, and when he was forced into a "dynamical" discourse he used such pre-Newtonian concepts as "violent" or "natural" motion, which did not satisfy Mersenne and probably did not satisfy even Torricelli himself.[40]

In February 1645, Mersenne let Torricelli know that Descartes and Roberval, too, had doubts about Galileo's science of motion.[41] The dispute peaked in January 1646, when Roberval wrote Torricelli a long letter dealing mainly with pure geometry and initiating the controversy over the cycloid, but also criticizing Galileo's theory of motion.[42] Roberval claimed that falling bodies behave as Galileo said only at the beginning of their fall and later attain a constant maximum speed. Empirically, considering the effect of air resistance, Roberval was right. But Galileo's law of free fall assumed an ideal geometrical case, or falling in a vacuum, while Roberval spoke of bodies falling in air.

Torricelli could have resolved the issue by telling Roberval to read Galileo's *Two New Sciences* more carefully, which Torricelli did, but only implicitly. Amazingly, he chose to reply in a completely different way. First, he wrote to Ricci that he intended to emphasize his interest only in the mathematical theory rather than its physical application:

> I do not care whether the principles of the *De motu* are true or false. For if they are not true, let us feign they are true, as we have assumed, and then look at other speculations derived from these principles, not as mixed but as purely geometrical. I feign or suppose that a body or a point moves downward or upward with the known proportion and horizontally with uniform motion. If so, I say that it will conform with what Galileo and I, after him, have said. If then, lead, or iron or stone balls do not comply with these suppositions, too bad: we shall disregard them [the balls].[43]

To Roberval, Torricelli told the following story (quoting, as he said, from Galileo): Archimedes once supposed that projectiles move in spirals and hence composed a full treatise concerning spirals. Later he discovered that he was wrong, but did not throw his treatise away and only deleted the word "projectiles," leaving a mathematical treatise dealing with "general" points moving according to a *fictitious* law, rather than the *natural* one. Geometry, Torricelli remarked, relies only on itself, and is not affected by eventual disagreement with empirical facts; therefore he proposed to "reject all that is physical in that book, namely terms like ballistics, etc. and leave what is geometrical, namely abstract propositions. The rest is fable."[44]

Are we to understand that if Torricelli had had to rewrite his book on geometry he would have avoided speaking of the possible application of his mathematics? Did Torricelli, at this point, change his mind about Galileo's physics? His responses to more attacks a year later on Galileo's theory of projectiles as expounded in Torricelli's "De motu" may help to clarify Torricelli's position.

Renieri's Complaint

Three years after publication of Galileo's *Opera geometrica,* Giovanni Battista Renieri of Genoa, brother of Vincenzio Renieri, a professor of mathematics at the University of Pisa, wrote to Torricelli complaining that some experiments he had done with guns contradicted Galileo's theory. In his letter, dated August 2, 1647 he said: "Your work on the motion of projectiles, which reveals your very sharp intellect, has reached Genoa and given our gentlemen the opportunity to make several experiments with the shooting of various kinds of guns and I was astonished that such a well-grounded theory turned out to work so badly in practice."[45] The "gentlemen" Renieri referred to may have included Baliani who in the same year became governor of the fortress of Savona, west of Genoa.

Renieri said he had set a 2-ell (about 1.2 meters)-high gun at an elevation of 45 degrees (6 points) and found its maximum range to be 2,300 paces (a pace approximately equaled an ell). He then shot point-blank (zero elevation) and achieved ranges of 400 paces and more, much longer than the theory predicted. Renieri did not say what the predicted result was, or how he calculated it. The calculation is easy using the gravitational acceleration constant *g*, but *g* was still unknown in 1647. (I performed the calculation relying only on theorems known then and found an expected range of approximately 96 paces, about a quarter of that measured by Renieri.)[46]

Renieri concluded: "If not for the authority of Galileo, of whom I am a partisan, I would certainly feel some doubt concerning the motion of projectiles: whether it is parabolic or not, and if it is so, I would doubt that the axis of the said parabola is perpendicular to the horizon."[47] He then proposed an ad hoc variation of Galileo's theory, suggesting that the axis of the parabolic trajectory goes not through the center of the earth, as Galileo said, but slightly off center. Renieri may have been wrong but was certainly well versed in mathematics.

A few days later, presumably on August 8, Torricelli replied with a long, remarkable letter that began:

> When writing that small book on motion I never intended to make any assertion which is not *ex hypothesi,* in contrast to those who take for true

> those two very well-known suppositions: the spaces fallen by the body in equal times are *ut numeri impares ab unitate;* and that the spaces traveled horizontally in equal times are equal.[48]

Torricelli repeated that he wrote on hypothetical subjects and did not pretend, as some may have perhaps thought, to describe reality; both Galileo and he "speak geometrically."

Next, however, Torricelli wrote of the practical applications of the theory. He was well aware of possible sources of error:

> Many of these causes which can make experiments disagree with theory have been noticed by Galileo in his book on motion. The main one, however, is air resistance which resists any kind of motion but especially when the body is faster; and on the whole it will resist very strongly when the projectile is thrust by the supernatural fury of fire which is the highest of all our natural and artificial speeds. No wonder, then, if experiments, and mainly those made with fire machines, turn out different from theory. Our supposition is that the horizontal impetus remains unvaried; experience, however, shows that the horizontal impetus near the muzzle is four or six times greater than at the end of the shot.[49]

Nevertheless, Torricelli admitted that, even accounting for these errors, Renieri's results were too far from theoretical expectations. He conjectured that the discrepancy may have been caused by three factors: (1) The gun had an imperceptible elevation. (2) The plane of fire was not horizontal, that is, the gun's site and the landing point were not at the same height. (3) At the moment of shooting, the gun tilted up, thus lengthening the trajectory.

Torricelli proposed precautions to eliminate the errors: (1) Level the gun with several squares; slight imprecision in any one square may alter the range considerably. (2) Shoot on the seashore, or on a level plane to ensure that the gun and the landing site are at the same height. (3) Not knowing how to prevent inadvertent tilting of the gun, Torricelli suggested ways to ensure that the shot is indeed at point-blank: (a) In front of the gun place a square frame made of four canes, each two or three ells long, with paper stretched over it. The frame should be located so that the ball passes through the paper, providing a check point for tracing the trajectory. (b) Drop a cannon ball from the same height as the muzzle at the instant of firing to see whether it touches the ground simultaneously with the projectile.

Torricelli here displayed his experimental skill, when he might easily have said merely that guns are imperfect instruments. As Rupert Hall rightly points out about the state of the art in Torricelli's day, "The standard of engineering technology was not merely insufficient to make scientific gunnery possible, it deprived ballistics of all experimental foundation and almost of the status of an applied science, since there was no technique to which it could, in fact, be applied."[50]

Many factors other than air resistance rendered seventeenth-century guns imperfect. For example, gun barrels were irregular (grooved barrels had not yet been invented), easily distorted by the heat of firing and friction. Trunnions supporting the gun barrel on each side were not always symmetrical about the axis of the barrel, thus altering the angle of elevation. Elevation could also be affected, as Torricelli noticed, by tilting or vibration of the gun at the moment of firing. The quantity of gunpowder in a charge was not accurately measured, and its quality was not consistent, so that the thrust given to the ball, and consequently the muzzle velocity varied from shot to shot. Balls were not uniform in shape or density, so that the effect of air resistance also differed from one shot to another. Lastly, atmospheric factors, like air pressure (which is proportional to density), greatly influence the trajectory of projectiles but were not taken into account. Seventeenth-century science was unable to cope with these sources of error, and the experience and judgment of the gunner were the most important element in accurate shooting.

Torricelli's experimental instructions were not really very helpful, but Renieri spared no efforts to carry them out. In a second letter, written on August 24, Renieri described two sets of experiments he performed with a 3.5-palm-high falconet (a light cannon) having a point-blank range of 700 palms. (A palm was about 24.5 cm, so the falconet was approximately 86 cm, or 2.8 ft, high, with a range of some 171.5 meters, or 563 ft.) Renieri placed three paper-covered frames at distances of 100, 300, and 500 palms from the falconet. In theory the ball should have penetrated the frames at heights of about 3.4, 2.9, and 1.7 palms, respectively. In practice it struck at 1.5, 1, and a little below 1 palm, respectively. As the diagram (fig. 5.5) shows (K, L, and M are the points at which the ball passed through the frames), the trajectory was not only not parabolic, but part of it reversed curvature and had a shallow downward bulge. Renieri then raised the elevation to one minute (1/12 of a point), equivalent to 0.625 degrees, and got a range of 1,800 palms instead of the theoretically expected 1,530.[51]

Torricelli answered Renieri at the beginning of September, again stating the hypothetical principles of ballistics and repeatedly stressing that, deliberately to avoid controversies, his book on motion had been addressed to philosophers rather than gunners. As for Renieri's results, Torricelli illustrated theoretically that a downwardly concave

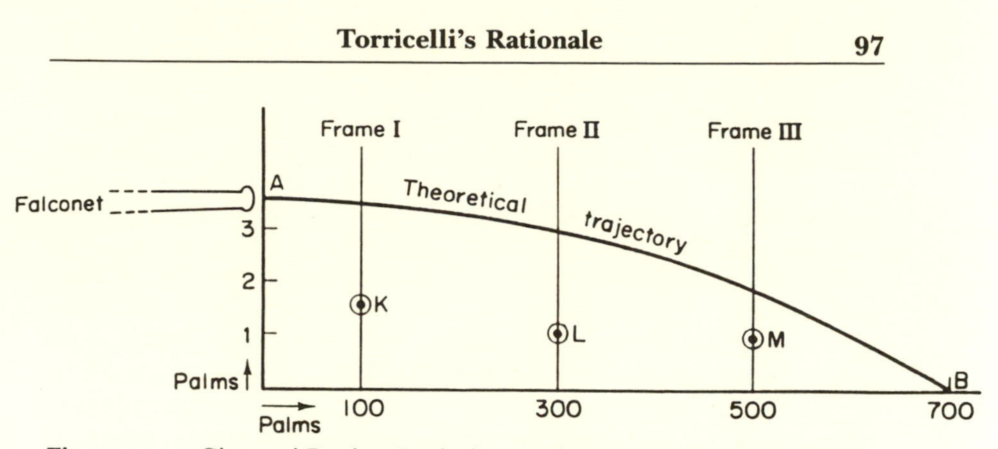

Figure 5.5. Giovanni Battista Renieri's experiment (reprinted with permission from *Annals of Science*). AB is the theoretical parabolic trajectory; K, L, and M are the points actually observed along the ball's path where it passed through paper-covered frames.

trajectory is absurd.[52] From a theoretical point of view Torricelli was right but in practice a trajectory like the one Renieri observed might be due to an uneven or oddly rotating cannonball. Baseball pitchers use skill in trying to achieve such a trajectory intentionally. However, Torricelli's knowledge of physics was inadequate to anticipate such effects.

The exchange of letters between Renieri and Torricelli then stopped. The rainy season prevented Renieri from carrying out additional experiments, and Torricelli died in October. Nevertheless, the correspondence shows that Torricelli, a year after Roberval's criticism, had not changed his mind about the soundness of Galileo's physics and was still convinced of its practical applications, but considered such applications to have only secondary importance. Had he changed his mind, he would not have bothered to give so many empirical instructions. What, then does this say about Torricelli's scientific approach?

Torricelli's Scientific Approach

Torricelli's approach was undoubtedly hypothetical-deductive, but his view of the role of experiment in physics remains ambiguous. Although he said that experimental results are worthless, he kept suggesting sophisticated experiments. Historians have therefore differed in their conclusions. Rupert Hall blames Torricelli for reopening the breach between scientific speculation and physical reality that Galileo had attempted to close.[53] Lanfranco Belloni calls Torricelli basically an instrumentalist, or "Bellarminist" (Belloni's term, referring to Cardinal Bellarmine), and even argues that Torricelli was not an experimenter but studied only from books, had a speculative approach to science, and helped Galileo in Arcetri mainly in mathematical formulations.[54]

But an "instrumentalist" is a scientist who cares only about the practical application of a science, whether a theory "works" rather than, say, whether it describes a real process. And both Belloni and Hall admit that on other occasions, for example, lens manufacturing and giving instructions to gunners, Torricelli did try to apply mathematics to physical reality. Hall remarks: "It was pardonable for the reader to suppose that when he talked of guns he meant real guns, that when printing tables giving measurements in paces he was not computing lines on a diagram, that when he designed a gunner's quadrant it was not to be used in the geometrical barrels of hypothetical guns." Furthermore, Torricelli did after all perform, or at least plan, the experiment that took his name. Why was he so ambivalent?

It may have had to do with religion, as Belloni hints in his designation "Bellarminist." Torricelli and his Medici patrons probably interpreted Galileo's condemnation by the Church as more than merely a taboo on Copernicanism but a warning to all men of science that the Church would not tolerate a claim to truth for any hypothesis, whether just conjecture or a mathematical model. To be on the safe side, they—like the Jesuits—may have adopted a general instrumentalistic attitude.

Although this interpretation may appear extreme, it is consistent with the behavior of the Medici and Torricelli. If the grand duke appointed Torricelli Court Mathematician and not Court Mathematician and Philosopher, he may well have been signifying that he was separating mathematics from philosophy, and hence from physics. Torricelli, for his part, was very cautious. He concentrated on geometry, wrote very little on physics, and nothing on astronomy. He did not even publish the results of his barometer experiment, despite its importance. The prudent phrasing of his replies to Renieri, emphasizing that he spoke *ex hypothesi* and that Galileo's laws of motion were only a mathematical speculation that could not "dictate" truth, can be understood in the same way.

Nevertheless, this explanation is only conjecture. As I have already argued, it is difficult to judge how much Torricelli, or even the Medici, had—in this context at least—reason to fear the Church. Galileo's formal condemnation, after all, concerned only the Copernican system, and no evidence, despite Redondi's claims, shows that Torricelli or the

Medici interpreted it more broadly and extended it to physics in general. As to Torricelli's title, assuming it was solely that of Mathematician, which has still to be proved, we are still not certain it reflected religious restraint. As to Torricelli's ambivalence about experimentation, it may have had other causes.

Paolo Galluzzi proposes a different explanation, that Torricelli was not really an instrumentalist but tactically pretended to be one to protect Galileo's work from justified criticism, which was becoming strong and unrefutable. Mersenne, Descartes, Roberval, and even Nardi, a Galilean himself, were presenting more and more experimental evidence undermining Galileo's theories. Eventually, the opposition reached a point when Torricelli could no longer find answers, although he remained convinced that Galileo was right. Unable to reject contrary empirical evidence, he chose to portray Galileo's science as generally detached from reality.[55]

In the light of the above considerations, Galluzzi's explanation seems more plausible. Yet, in my opinion, Torricelli's behavior can also be explained if seen from his own point of view, that of a mathematician. Torricelli, unlike Galileo, always regarded himself as a mathematician. In a 1632 letter, at the very beginning of his career, he presented himself to Galileo as a mathematician by profession and later wrote almost exclusively on mathematics; "De motu" was part of his geometrical work.[56] He appears to have shown an interest in physics only when it touched on his mathematics, and so probably felt no strong obligation to defend physical results. His examples, from the start, referred to *geometrical* rather than *real* space.[57]

It is sometimes said that one of Galileo's main contributions to science was the introduction of mathematics, geometry in particular, into physics. Torricelli here did the opposite, included physics—a "philosophical" domain—among the mathematical sciences, which constituted a distinct field of knowledge. Was Galileo trying to do the same thing? One cannot, of course, "transfer" this conclusion from Torricelli to Galileo. One can say, however, that Torricelli, although he may have been an experimenter, was certainly not an experimentalist, that is, did not use experiment as a heuristic device, a basis for scientific discovery. This is particularly clear when Torricelli presented and defended Galileo's work.

Throughout we have seen that it was Galileo's critics rather than he and his supporters who engaged in empirical discourse. How, then, did he acquire his popularly accepted reputation as an empiricist scientist? The origins of this image are to be found in Galileo's biography by Viviani.

6

Viviani's Hesitations

This short chapter and the one that follows deal with the decade after Torricelli's death, when little or no science was practiced in Tuscany. The general picture of Galilean science in this period is rather gloomy. Galileo and most of his leading followers had died, Castelli and Berti in 1643 (only one year after Galileo), Torricelli, Vincenzio Renieri, and Cavalieri in 1647; all traces of Nardi in 1647 are lost, and he, too, probably died in that year. Finally, Galileo's son Vincenzio, who had collaborated with his father and with his father's followers, died in 1649. Thus only Vincenzio Viviani remained in Tuscany in the 1650s to pursue Galileo's work. His main contribution, especially in the decade after Torricelli's death, was to document the history of Galilean science, rather than to add to it. This period saw the collection of Galileo's manuscripts and their first historical description. To understand how this was done and Viviani's contribution to the history of science, it is helpful briefly to review Viviani's life and personality.

Viviani

Vincenzio Viviani was a Florentine of noble birth associated with the Tuscan court throughout his life. Like Torricelli, he received his basic education from the Jesuits and studied logic under the Franciscan father Sebastiano da Pietrasanta, confessor of Prince Leopold de' Medici. Father Sebastiano encouraged Viviani to study geometry, and the boy joined the exclusive circle of students of Clemente Settimi, a mathematician monk of the Pious Schools who collaborated with Galileo. Viviani was a child prodigy, and when he was presented at court at the age of sixteen, the grand duke was so impressed with his mathematical talent that he awarded him a scholarship and introduced him to Galileo.[1]

Galileo, then seventy-four years old, blind, and confined by the Inquisition to his house in Arcetri, was still very active as a scientist. His *Two New Sciences* appeared in that year (1638), and he was still engaged in many scientific projects. However, he needed assistance and welcomed Viviani as his amanuensis; in 1639 Viviani moved into Galileo's house to work with him. As Viviani himself said, his relations with Galileo were not confined to scientific collaboration but assumed the character of father and son.

In 1641 Torricelli left Rome to join Galileo and Viviani in Arcetri. The two young mathematicians became good friends. After Galileo's death, in 1642, Torricelli was appointed Tuscan Court Mathematician in Galileo's place and lecturer in mathematics at the Studium of Florence. Viviani continued to collaborate with him and in 1644 helped with the barometer experiment. After Torricelli's death, in 1647, Viviani took his place as lecturer at the Accademia del Disegno—the first step in a long and successful intellectual career.

The story of Viviani's career is related in a series of impressive documents collected in Gal. MS 155. In 1649 he began tutoring the pages of the grand duke, normally a duty of the Court Mathematician, and was clearly regarded as Galileo's and Torricelli's successor at court. In 1653 he became "Ingegnere della Parte Guelfa" (Engineer of the Guelph Party), in charge of the Office of River Control with the rather low initial salary of only seven scudi per month.[2] Four years later, in 1657, he was inducted by Prince Leopold into the most prestigious scientific society of those days, the Accademia del Cimento, and also became a member of the equally prestigious literary Accademia della Crusca. In 1664 he reached the peak of his fame: Together with twelve other leading men of letters in Europe, he was granted a special pension by the French king Louis XIV. The honor was great, though the pension was small. But Viviani's many public engagements, especially as court engineer, hindered his research work, and, as he himself admitted, he was not very successful as an engineer because his duties required much riding and more physical strength than he had.[3] In 1666 the grand duke relieved him of his public obligations and gave him a salary sufficient to enable him to concentrate on his studies.[4]

When Viviani retired he had more time to continue gathering laurels. In the late 1660s he was considered for the prestigious chair of mathematics at the University of Padua, held half a century earlier by Galileo. The University of Padua normally offered high salaries to outstanding foreign mathematicians but was then experiencing financial restrictions and could not offer the salary Viviani deserved.

Viviani did not get the appointment; either the university withdrew his name, or he declined an unsatisfactory bid.[5] He was also offered high scientific positions in France and Poland but declined them, choosing to remain in Florence, eventually at the disposal of the Tuscan court. Viviani's nephew, Jacopo Panzanini, wrote a short, still unpublished biography of his uncle, listing several official missions Viviani filled abroad in his court capacity.[6] Later, in 1690, when the prominent literary Accademia Arcadia was founded, he became a member under the pseudonym of Erone Geonio.[7] In 1696 he became a foreign member of the Royal Society of London and, three years later, a foreign member of the Académie Royale des Sciences in Paris. He died in 1703 at the age of eighty-one.

Historians praise Viviani's scientific work highly, and I presume they do so to compensate for ignoring it, since he published relatively little and neither his published writings nor his unpublished manuscripts have received much study. The amazing fact is that these Viviani manuscripts fill 104 volumes in the Galilean collection, about the same number as those documenting the lives of Galileo and Torricelli together.[8] This is enough to discourage the most enthusiastic researcher; even scholars with the perseverance of Favaro and Caverni—who praised Viviani—avoided delving into this mass of materials.

However, not everybody esteemed Viviani. Several scientists who worked with him, among them Italy's leading scientist, Borelli, who moved to Tuscany in 1656, did not think highly of him.[9] Borelli's reasons may have been personal, but it is true nonetheless that Viviani published little, and nothing outstanding, mostly restorations of classical works in the tradition of Renaissance mathematics. They do reveal a mathematical talent but hardly put him on the level of Galileo, Cavalieri, or Torricelli. Viviani's *De maximis et minimis* (1659), for instance, was a good restoration of Apollonius's fifth book of *Conics,* which we have today but which at that time was not available and was believed to be lost.[10] But even this work proved to have little scientific value because, unfortunately for Viviani, shortly before its publication Borelli found an Arabic version of the original in the library of the grand duke and arranged its translation into Latin. At the court's request, Borelli withheld the translation until Viviani could complete his restoration.[11] Viviani's later works included some collections of mathematical problems, a short treatise on floods, and a divination of some of the lost works of Aristaeus the Elder, all relatively minor.[12] Perhaps more important writings may one day surface from the vast number of his unpublished manuscripts.

Why did Viviani publish so little of what he wrote? How did he succeed in making such a brilliant career despite the slightness of his few published works?

A. Natucci, author of the article on Viviani in the *Dictionary of Scientific Biography* (1976), claims that Viviani turned his talents and ingenuity solely to the study and imitation of the ancients because the Church prevented him from pursuing the mathematical ideas evolving during that period.[13] Yet, as I will point out in detail later, I found little evidence that the Church interfered with the activities of either Viviani or Galileo's other followers, except for Viviani's teacher Clemente Settimi. Settimi was briefly imprisoned in 1641 by the Roman Inquisition on suspicion of having read Galileo's forbidden *Dialogue.*[14] Viviani, who in 1654 was refused permission to read the *Dialogue,* was certainly cautious.[15] But this hardly signified severe Church interference; as his correspondence testifies, he could readily have followed the evolution of mathematics in his day. He received letters from all over Europe, including Protestant countries, from prominent cities such as Rome, Venice, Paris, London, and Prague. He also met or knew personally many leading contemporary scientists, such as Malpighi, Leibniz, and Robert Southwell (later president of the Royal Society of London).[16] My reading of Viviani's manuscripts indicates rather that he was a very hesitant man and that it was his habitual hesitance, not necessarily intimidation by the Inquisition, that prevented him from completing work he had started.

Viviani was also a perfectionist and, although this was to some extent a characteristic of his era, in Viviani it apparently amounted to a neurosis (to the benefit—as we shall see—of meticulousness in the history of science). His perfectionism is evident, for example, in the many preliminary drafts of some of his writings and in his long and detailed letters. Viviani may have had what the modern philosopher of science I. C. Jarvie calls a "thoroughness mentality," which holds that scholarship demands an exceedingly comprehensive knowledge of the most minute elements of a subject.[17] This kind of mentality certainly hindered Viviani's efforts and was the main cause—Freud would probably say excuse—for his reluctance to publish. He was always hesitant and rarely completed what he started; even after the grand duke freed him from obligations and granted ideal working conditions, Viviani produced but little. In his correspondence, he often blamed poor health as a debilitating hindrance. Maria Luisa Righini Bonelli, one of the few modern historians who has made use of Viviani's manuscripts, has a different explanation. In her view

Viviani's mind was not capable of coordinating what he grasped, so almost all his attempts at scientific achievement were tentative and unfruitful.[18]

Viviani's success in life suggests that standards for evaluating merit have changed little from Viviani's day to ours; he was a refined, educated man and a skilled courtier who knew how to use good manners and public relations for personal advancement. His pension from Louis XIV, for instance, may have signified more the king's self-glorification than a recognition of intellectual excellence. Hints of this appear in a letter from Louis's minister, Colbert, to Viviani in 1664: "And if you will strive to give the public something of the glory of such a great prince [Louis XIV], although you may be pledged to this by any other motive than self-interest, it will be a certain means of obliging him to continue [the payments] in the future."[19] And, to paraphrase W. E. K. Middleton, Viviani also made a good deal of capital out of having been Galileo's last pupil, perhaps more than he made out of his own scientific work.[20] He was not unique in this respect among Galileo's followers. As seen in previous chapters, some relatively mediocre Galileans like Niccolò Aggiunti and Dino Peri received the chair of mathematics in Pisa in preference to excellent candidates like Cavalieri and Borelli.

Although Viviani's own scientific achievements were slight, his devotion to Galileo was outstanding. Shortly after Galileo's death Viviani, not yet twenty, was already planning to collect and publish the master's works and manuscripts.[21] From then on he dedicated his entire life to searching for and amassing material relating to Galileo and his work, which is how the Galilean collection came into being. Viviani's devotion to Galileo's memory sometimes appears excessive, even obsessive. He gathered as many of Galileo's papers and belongings as he could find, made endless inquiries about the details of Galileo's life, placed a bust of Galileo above the entrance to his home accompanied by an inscription in Galileo's memory, and even asked to be buried near him.[22]

Galileo's Collected Works

When Torricelli died, his and Galileo's published works contained only part of their scientific writings. Of Galileo's two main books, the *Dialogue* and the *Two New Sciences,* only the latter was available in Italy, and since it had been printed abroad, Galileo's remaining Italian followers encountered difficulties distributing and eventually reprinting it. Most of Torricelli's works, including his description of the barometer experiment, remained unpublished. Meanwhile, the barometer experiment was being repeated in various parts of Europe, and Valeriano Magni, a well-known Italian-born Franciscan mathematician in Poland, claimed priority for it. Also, as noted before, Pascal in France had questioned Torricelli's priority for the theory of the cycloid.

Galileo's and Torricelli's work were thus in danger of being misunderstood, misrepresented, and, worse, discredited. Any depreciation of their accomplishments could by association reflect adversely on the patronage of the Medici. Such apprehensions may have spurred the urge, evident in the correspondence of Galileo's followers, to collect Galileo's and Torricelli's writings before it was too late.

Efforts to collect, order, and publish Torricelli's manuscripts were unsuccessful, but more luck was had with Galileo's.

The Medici were at least as eager as Viviani to see Galileo's papers gathered and reprinted. In 1650 Prince Leopold de' Medici wrote a confidential letter to one Virgilio Spada, inquiring about the possibility of receiving permission to republish Galileo's works, even if this entailed a concession to the Inquisition in the form of a few omissions and modifications.[23] In 1655 the Medici and Viviani welcomed the initiative of Carlo Manolessi, a publisher in Bologna, to collect and publish Galileo's works, except, of course, the forbidden *Dialogue*. They sent him the Galilean papers they already had, searched for Galilean material all over Europe, and Prince Leopold asked the Tuscan Ambassador in France to look for Galilean manuscripts in France and Holland. Manolessi also asked Prince Leopold several times for help in persuading the Inquisition to be more lenient in granting permission to publish some controversial passages.[24]

As part of this effort, Prince Leopold asked both Viviani and Gherardini to write sketches of Galileo's life. As Viviani reported, the prince had in mind to ask a "very literate" person to write a final version of Galileo's life for inclusion in Manolessi's edition.[25]

However, things did not turn out as Galileo's patrons and followers had hoped. The Bologna edition appeared in 1655–1656 without the life of Galileo, probably because the Manolessi product did not satisfy Galileo's followers; Viviani on one occasion complained that Galileo's works deserved a more elegant presentation.[26] The final biography of Galileo may never have been written, and if it was, it did not survive.

Viviani's and Gherardini's outlines, however, did survive, and we owe their preservation chiefly to Viviani, along with that of many other Galilean documents. Viviani contemplated a grand edition of Galileo's works, with Latin translated into Italian and vice versa, and amassed an enormous quantity of material, which is now a large part of the Galilean Collection of manuscripts in the National Library of Florence. He may also have himself translated, or had translated, some of Galileo's works; for instance, Gal. MS 316 contains Latin translations of some of Galileo's Italian writings.

But Viviani failed to finish his ambitious project, for the same reason he published so few of his own works; he was ever the perfectionist, never satisfied that the material he had gathered was comprehensive enough, and reluctant to stop collecting and to begin publishing. As early as 1656, Viviani already had on hand enough Galilean material for an outstanding edition, even within the limitations set by the Inquisition, but may have thought he must first make sure he had obtained and read every last bit that existed. An edition of Galileo's works, as Viviani would have liked to see it, did not appear until two centuries after Viviani died. However, Antonio Favaro, who supervised the National Edition, could not have accomplished his immense task without the papers collected by Viviani.[27]

Viviani's life of Galileo, the "Racconto istorico," was one of the most important seventeenth-century contributions to Galilean studies. This document, treated in detail in the next chapter, was probably the earliest biography of Galileo, certainly the earliest extensive one.[28] Part of its importance is that it was the source and inspiration of the trend in science history to portray Galileo as an empiricist and, more generally, of the "Baconian" approach, which emphasizes the empirical aspects of the scientific revolution.

7

Patterns of a Renaissance Biography

The "Racconto Istorico"

We now return to Galilean historiography. I will try to show how, in writing the "Racconto istorico," his life of Galileo, Viviani conformed to the biographical standards of his day—which tended to heroize geniuses—and perhaps also anticipated his readers' lack of interest in theory and mathematics. His portrayal was fitting for that time but not reliable for the purpose of a modern history of science. Its amplification into the Galileo myth was due primarily to distortions by later biographers.

The "Racconto istorico" was eloquent and concise, now filling no more than thirty-four pages in the National Edition. Even a Galileo expert has difficulty in detecting its inexactitudes; it not only flows smoothly but carries the distracting cachet of long-accepted truth. Viviani covered all the stages of Galileo's life and career: his youth and studies in Florence and Pisa, professorships in Pisa and Padua, many controversies, telescopic discoveries, return to the Tuscan court, campaign for Copernicanism, trial, and, finally, scientific work in old age, when Viviani collaborated with him. Viviani ended with general remarks about Galileo's personal traits—cheerful, open-minded, generous, easily angered but quick to forgive, and well versed in arts and letters—and also listed some of Galileo's friends and followers.

Many of the details reported by Viviani, in addition to those already mentioned in chapter 2, are in doubt. Viviani averred, for instance, that Galileo had to leave Pisa partly because an unfavorable opinion he expressed on a project to dredge the port of Leghorn offended a high-ranking personage. There is no trace of such a project in the Tuscan archives.[1] Viviani also claimed that one of Galileo's students in Padua was none other than King Gustavus II Adolphus of Sweden. There is no evidence that Gustavus Adolphus ever visited Italy.[2]

Emil Wohlwill, in 1903, was the first to question the validity of these stories.[3]

Nevertheless, many of Viviani's details were confirmed by Gherardini in his own sketch of Galileo's life, probably written independently of Viviani since Viviani annotated and corrected it (Favaro published Viviani's annotations together with Gherardini's biographical sketch in the National Edition). Gherardini described Galileo's life only until his return from Padua to Florence in 1610, but repeated Viviani's story about Galileo's objections to the port-of-Leghorn project, and even identified the offended personage as Giovanni de' Medici, the illegitimate son of Grand Duke Cosimo I.[4] Since Viviani was in the Medicis' service, he may diplomatically have avoided naming Giovanni. Gherardini also placed Gustavus Adolphus among Galileo's students in Padua,[5] and, in general, his image of Galileo was similar to Viviani's.

To understand Viviani's and Gherardini's accounts in the context of the era in which they wrote them, it may be helpful to look at their various handwritten drafts and at some related documents that Viviani diligently collected (now in the Galilean collection of manuscripts). Most are in Galilean MS 11, and many are still unpublished.

Galilean MS 11

Galilean MS 11 contains two drafts of Viviani's "Racconto istorico," a copy of Gherardini's life of Galileo with Viviani's annotations, a few undated biographical notes on Galileo by his son Vincenzio, and many other documents, some written or annotated by Viviani, such as genealogical data on Galileo's family.[6] In 1659 Viviani also wrote a second essay about Galileo's work—"Lettera di Vincenzio Viviani al Principe Leopoldo de' Medici intorno all'applicazione del pendolo all'orologio"—presenting his version of Galileo's discovery of the pendulum principle.[7] Of the four early biographical essays on Galileo in volume 19 of the National Edition, Viviani's was the first to be published and the most important.

Favaro, editor of the National Edition, labeled the two drafts of Viviani's "Racconto" in Galilean MS 11 *A* and *B;* the earliest (posthumously) printed version of 1717, which Favaro labeled *S* after the name of its editor, Salvino Salvini, differs slightly from *A* and *B* and is therefore also important.[8] All these versions bear the date April 29, 1654, but *S* was an improvement of *B,* and *B* was an improvement of the earliest version, *A,* which may have been written for the Manolessi edition (an unproven surmise). The changing handwriting indicates that Viviani wrote *A* and *B* at different periods of his life; *B,* the later, was almost certainly the version he was preparing for his planned edition of Galileo's works.

Favaro considered publishing all three versions in the National Edition superfluous, since their differences seemingly added nothing to an understanding of Galileo. Favaro, well aware that Viviani had been a perfectionist, remarked that the many small variations between drafts exemplified Viviani's continuous and endless quest for the ideal text. For instance, in 1668 Viviani had written to Michelangelo Ricci that his life of Galileo was "approaching perfection"; twenty-four years later in 1692, when Viviani was seventy, a letter to Domenico Soderini (Gal. MS 11, 174) shows he was still inquiring about Galileo's date of baptism, to complete his documentation. Hence Favaro chose to publish only version *B,* because it was the latest authentic text, and gave in footnotes the variations among the three he considered worth noting.[9]

During nearly a century since Favaro's lifetime the history of science in general and Galilean studies in particular have progressed considerably. Some corrections in Viviani's various drafts now appear illuminating rather than trivial. My reading, for instance, shows that Viviani's "corrections" often correspond exactly with suspected distortions. Another striking aspect of both Viviani's and Gherardini's drafts is the many alterations that were stylistic rather than substantive, which tells us, among other things, the type of readers they had in mind.

Viviani's Audience

As might be expected of a perfectionist and a Florentine writing during a period when learned people were tending to adopt the Italian (i.e., Tuscan) language, Viviani's prose was polished. Yet his drafts, especially *B,* contained so many purely stylistic revisions that he likely was aiming his work at a specific audience with elevated literary tastes, and not merely other scientists. Draft *A* was formally addressed as a letter to Prince Leopold, but Viviani obviously had a larger audience in mind. *B* had two different interchangeable beginnings—both published by Favaro. The first repeated the beginning of *A* and also had the form of a letter to Leopold; the second was a somewhat rhetorical introduction that turned the "letter" into a general essay, not addressed specifically to the prince.

Gherardini's drafts also contained many stylistic alterations, implying that he too was probably writing for the same highly literate audience. Although he may not have been as learned as Viviani, the style of his final product was as clear, if not as polished, as Viviani's.

Who were the readers Viviani (and Gherardini) were catering to? Although Viviani did not explicitly identify them, we do know that he intended to include his life of Galileo in his planned collection of Galileo's works, and that the collection was to have Italian translations of original Latin, and vice versa. The planned use of Italian and the pains Viviani took with elements of style indicate that his intended readership was broader than professional scientists. Scientist contemporaries of Viviani, such as Cavalieri, Torricelli, and Borelli, when writing exclusively for other scientists, almost always did so in Latin with little attention to style. Although a lack of elegance does not necessarily preclude clarity, obscurities in the writings of these three scientists have often frustrated historians of science.[10] In contrast, Viviani, like Giordano Bruno and Galileo, wrote in Italian for the general educated public, and his meticulousness about style reflected their standards.

The educated Italian public in those days consisted largely, but not exclusively, of the learned nobility and clergy, and, as Furio Diaz says in his history of the Medici grand duchy, nobles associated with the court predominated.[11] This public used to meet in the various literary academies, such as the Florentine Academy and the Accademia della Crusca, in which Viviani was active.

The academies were for Viviani, as they had sometimes been for Galileo, the best avenue to a broad intellectual public. For one thing, Viviani had free access to them, which he did not have to conservative centers of learning like the University of Pisa, the most important and prestigious cultural institution in Tuscany. Moreover, the academies had a broad and heterogeneous membership, ranging, in addition to the educated nobility, from rich merchants with artistic and literary tastes to academics—clerics and laymen—with interests beyond their specific occupations.

However, the literary academies had particular predilections and were not very fond of mathematics. A 1642 letter from Cavalieri to Torricelli, after Torricelli was admitted to the Accademia della Crusa, gave an interesting description of the preferences of academy members and, in general, of the educated public, as far as science was concerned. Cavalieri congratulated Torricelli and advised him on how to speak to a meeting of the illustrious academy. He wrote first about the members' expectations:

> I hear that they expect physical rather than mathematical things, and perhaps they are right, for the former resemble more the chaff [*crusca*], whereas the latter is the flour—the true food and nutriment of the intellect. It is advisable to meet their expectations, and more than that, the *universal* expectation, that has little esteem for mathematics, unless it sees some applications.[12]

Cavalieri then scornfully told Torricelli how to deal with this type of intellectuals:

> It is therefore advisable to have ready two types of argument to satisfy them all. And more than that, to satisfy the public, which decides the value of doctors and doctrines by the number of their followers, one has to avail oneself of what is more easily sold, so as to serve the public better and deceive it, or better, to kill the intellect, because the public wants to be treated this way.[13]

Cavalieri thus advised Torricelli to avoid mathematical topics. Torricelli followed the advice: His *Lezioni accademiche*—the lectures he delivered (in Italian, of course) at the Accademia della Crusca between July 1642 and September 1643, published posthumously in 1715—dealt mainly with physics. The presentation was descriptive, rhetorical, and free of mathematical formulations, very different from Torricelli's other writings, which were almost entirely devoted to geometry and, except for a few passages, written in Latin with relatively little rhetoric.

Cavalieri's letter throws considerable light on the audience for which Viviani was writing. Viviani may well have heard Torricelli's lectures and may even have read Cavalieri's letter, as it was part of the material Torricelli left him. In any case, Viviani's "Racconto istorico" on Galileo complied with Cavalieri's prescription: It used language impressively and persuasively, like Torricelli's *Lezioni accademiche*, detailed the empiricist aspects of Galileo's work, and described its applications. Viviani, one may assume, never imagined that some of his later readers would be incredulous historians of science.

Viviani was therefore almost certainly writing to fulfill the specific expectations of the literary academies. Furthermore, the general educated public of his day wanted to hear about the practical applications of science, not mathematics, and he had to adapt himself.[14] Yet the preferences of his audience were not his only constraint. Other textual revisions indicate that he also adapted his writing to the pattern that then prevailed for such biographies.

Vasari's Heritage

Lives of great figures, including scientists, are often distorted by their followers, so that *vite* ("lives") become "hagiographies." This was particularly so during the Renaissance, when history was still influenced by ancient classical customs, and biographies followed a traditional model.

Classical examples are Giorgio Vasari's famous *Vite,* a collection of biographies of Renaissance artists, published for the first time in 1550 and reissued in 1568 in an enlarged edition. Another example, related more directly to the history of science, is *Vite de' Mathematici,* by Bernardino Baldi (1553–1617) of Urbino. Baldi, once a pupil of the mathematician Federico Commandino (1509–1575), between 1587 and 1595 wrote no fewer than 202 lives of mathematicians and philosophers from antiquity to his own day, of which only 57 have so far been published.[15] Like Vasari, who wrote his *Vite* to honor his teacher Michelangelo, Baldi wrote his *Vite* to honor Commandino. P. L. Rose, who studied these biographies, says that Baldi may have taken Vasari as a model, but other models were also available, for example, Diogenes Laertius, the biographer of Greek philosophers.[16]

Was Viviani writing in the Vasarian manner? Without delving into the complex issue of the development of contemporary historiography, of art in particular, it is certain that in Viviani's day Vasari's *Vite* were still considered a model. In 1647 they were republished by none other than Galileo's editor in Bologna, Carlo Manolessi, awakening a new wave of interest all over Italy in the history of art. Torricelli's and Viviani's friend, Carlo Dati, for instance, in 1667 wrote his *Vite* of ancient artists. The history of art and biography, in general, became more critical and erudite: The new style is exemplified by the work of two mutually antagonistic historians of art, Carlo Cesare Malvasia (1616–1693) and Filippo Baldinucci (1626–1696).[17]

Malvasia was a Bolognese historian of art who wrote in the Vasarian style, although he criticized Vasari for, among other things, neglecting non-Tuscan artists, particularly those from Bologna.[18] I have found no evidence of any direct link between Malvasia and Viviani, but Malvasia, like Viviani, was active in various academies and, as the modern historian Martino Capucci points out, wrote for an audience, much like Viviani's, of "academics, men of letters and patriots."[19] In *Le pitture di Bologna,* written three years after Viviani wrote his "Racconto istorico" and published in 1686, Malvasia praised empirical work with expressions reminiscent of Viviani's. Speaking of his own work Malvasia said: "It would suffice for me to guide you where you might be persuaded only through *simple eye inspection.* The evidence of the fact should be the one that permits you to judge; as, by means of *experience,* it is done in remote England as well as in nearby Florence."[20] Malvasia's readers, in Bologna, seem indeed to have had the same expectations as Viviani's in Florence.

Although Viviani may have had no direct contact with Malvasia, he certainly did with Baldinucci. Baldinucci, in general, had much in common with both Vasari and Viviani. Like them, he was in the service of the Tuscan court and even shared Viviani's patron, Prince Leopold de' Medici, for whom he collected artistic works (we owe to Baldinucci an important part of the Uffizi collection); like Vasari before him, Baldinucci wrote a collection of lives of artists; and, like Viviani, Baldinucci was a member of the Accademia della Crusca. As an art historian and biographer he was still considerably influenced by Vasari, but was more thorough and precise—he was the first art historian to make full use of documents—and corrected many of Vasari's errors.[21]

Viviani's writing should be considered in the context of Vasari's style, with the modification of more scholarship and documentation. Indeed, the recto of folio 168 in Galilean MS 11 (fig. 7.1) contains a quotation from Vasari, in Viviani's own hand, related to Michelangelo's death, and its verso (fig. 7.2) contains a note inquiring about Michelangelo's death from Viviani to none other than Baldinucci. The reply (further down on the same page), probably written by Baldinucci, is another excerpt from Vasari. I will refer below to the exact use Viviani may have made of these excerpts.[22] As we shall soon see, Viviani's life of Galileo had, in general, much in common with, and in appearance could easily fit into, Vasari's *Vite,* but also like Baldinucci's biographies, was better documented than Vasari's.

The classical Renaissance practices in biographies of artists, including Vasari's, Malvasia's, and Baldinucci's, were described in a study carried out fifty years ago by Viennese scholars Ernst Kris and Otto Kurz. Kris and Kurz considered a large number of Renaissance biographies of artists from Giotto (who died in 1337) to Rembrandt (1606–1669) and found leitmotifs rooted in Greek mythology that tended to heroize the artist. These recurring elements had two prominent components: the youth of the artist and his supposed supernatural knowledge of nature.[23]

A common feature was an alleged link between the artist and some other great man. Giotto, for example, was presented by Dante in his *Divine Comedy* and by later biographers, mainly during the fifteenth

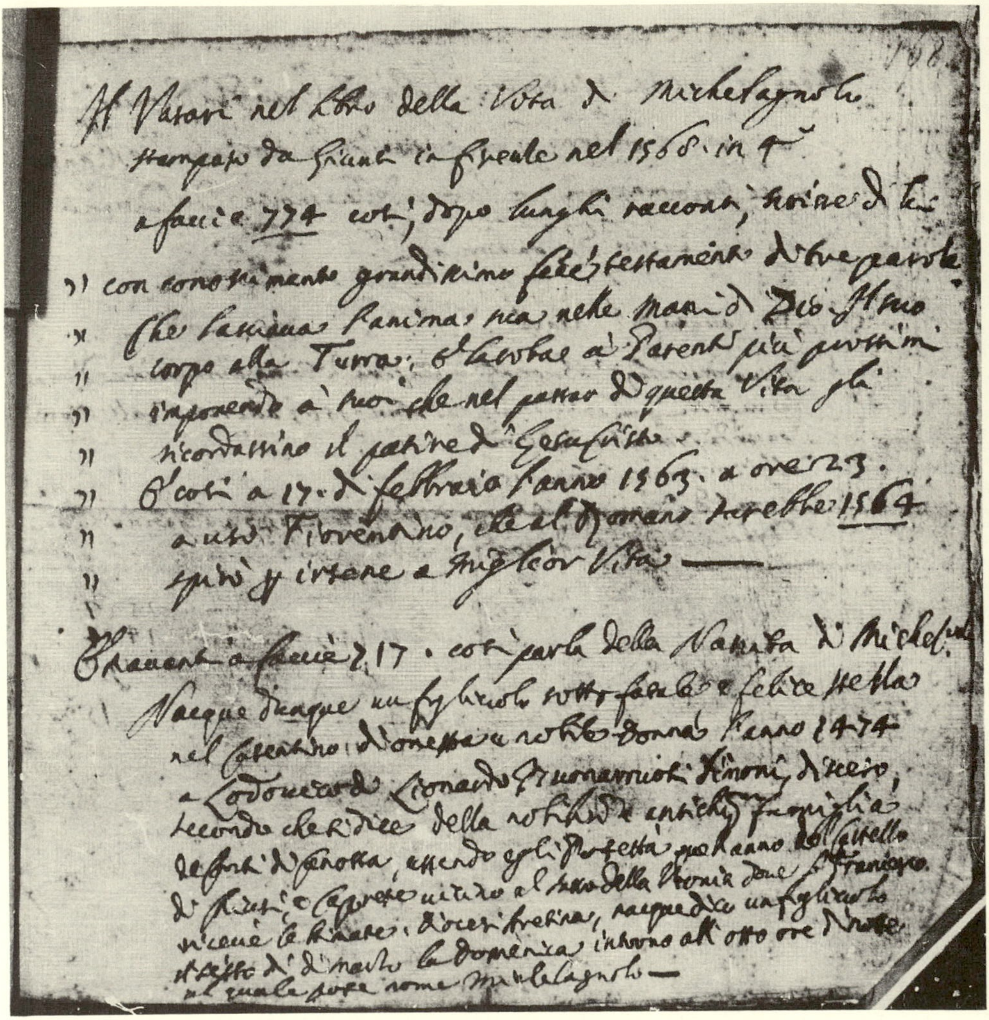
Il Vasari nel libro della Vita di Michelagnolo
stampato da Giunti in Firenze nel 1568. in 4°
a faccie 774 così; dopo lunghi racconti, soggiunge di lui:
" con conoscimento grandissimo fece testamento di tre parole
" che lasciava l'anima sua nelle mani di Dio, il suo
" corpo alla Terra, e la roba a' Parenti più prossimi
" imponendo a' suoi che nel passar di questa Vita gli
" ricordassino il patire di Gesu Cristo.
" E così a 17. di febbraio l'anno 1563. a ore 23.
" a uso Fiorentino, che al Romano sarebbe 1564
" spirò per irsene a miglior Vita —

Ed avanti a faccie 717. così parla della Nascita di Michel.
Nacque dunque un figliuolo sotto fatale e felice stella
nel Casentino di onesta e nobile donna l'anno 1474
a Lodovico di Lionardo Buonarroti Simoni, discese,
secondo che si dice, della nobilissima e antichissima famiglia
de' Conti di Canossa, essendo egli Podestà quell'anno del Castello
di Chiusi e Caprese vicino al Sasso della Vernia dove San Francesco
ricevé le Stimate: Diocesi Aretina, nacque dico un figliuolo
il sesto dì di marzo la domenica, intorno all'otto ore di notte
al quale pose nome Michelagnolo —

Figure 7.1. Galilean MS 11, folio 168r, with the excerpt of Vasari's life of Michelangelo in the top half. (Courtesy of the Biblioteca Nazionale Centrale, Florence)

century, as the personal pupil of Cimabue.[24] Although Giotto was certainly influenced by Cimabue, it is not at all certain that the two painters ever met. The noble or heroic figure of the great man was often depicted as a substitute for the real father of the artist—Cimabue for Giotto's simple peasant father. The artist's importance was also enhanced in other, similar ways: Native genius was sometimes said to

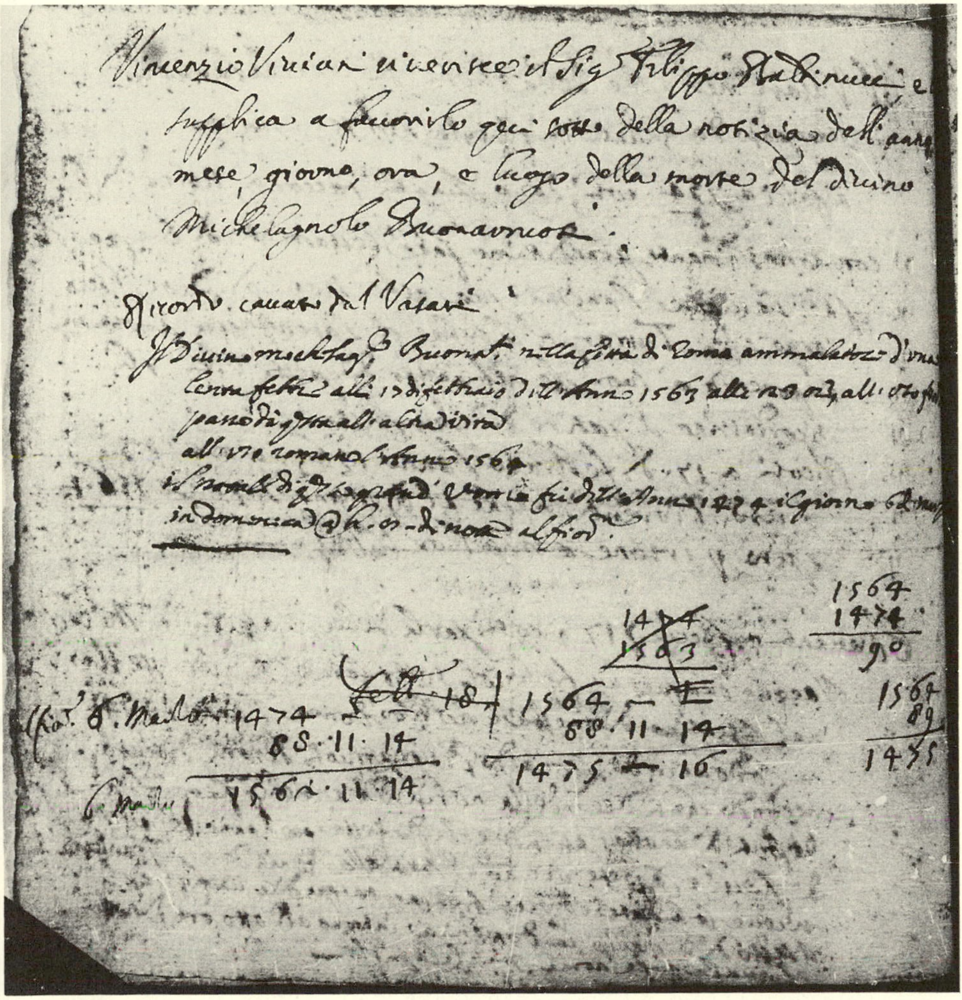
Vincenzio Viviani riverisce il Sig. Filippo Baldinucci e supplica a favorirlo qui sotto della notizia dell'anno, mese, giorno, ora, e luogo della morte del divino Michelagnolo Buonarroti.

Ricordo cavato dal Vasari

Figure 7.2. Galilean MS 11, folio 168v. Top: Viviani's note to Baldinucci. Center: A quotation from Vasari. Bottom: Viviani's calculations concerning Michelangelo's life (see also fig. 7.6).

have a divine origin, and the artist's birth was sometimes dwelt on as if the son of a god were being born. According to Kris and Kurz, Vasari's details about Michelangnelo's hour of birth and its precise astrological constellation were an example of this tendency.

Artists were often portrayed as child prodigies who commonly found their future vocations through coincidence. For instance,

Giotto—initially a simple shepherd—was said to have met Cimabue by pure chance while working as an apprentice to a wood dealer, and then ran away from the workshop in order to paint. The story was amplified by Ghiberti, followed by Vasari, so that Giotto became a shepherd drawing a sheep on a smooth piece of rock when Cimabue happened to pass by, recognized the lad's talent, and trained him to become a great artist. Vasari himself is known to have invented many anecdotes, including that of Giotto's famous ability to draw a perfect circle.

As these anecdotes show, the artist was believed to possess a knowledge of nature more profound than that of a layman, was capable, like a magician, of envisaging the whole from a single part. For example, when shown only a claw, he could determine the size of the lion.

Although Kris and Kurz limited themselves to biographies of artists, their conclusions apply equally well to biographies of scientists of the same period; Viviani's life of Galileo, at least, is a striking confirmation. His description of Galileo's youth, for instance, was typical of the "Vasari mode." The circumstance that Galileo was born on a date so near the death of Michelangelo, hero of the Renaissance, was an irresistible tempation for Viviani, who accordingly made the events coincide. Alas, Galileo was not born exactly on Michelangelo's date of death (February 18, 1564). When, then, was Galileo born?

Michelangelo's Death and Galileo's Birth

According to Salvini's version of Viviani's "Racconto istorico" (draft *S*), and to the *Dictionary of Scientific Biography* (1972, article by Stillman Drake), Galileo was born on February 15, 1564.[25] But Viviani's original version (drafts *A* and *B*) and the biographical notes of Galileo's son, Vincenzio Galilei, gave the date as February 19. Gal MS 11, 161—a genealogical tree of Galileo's family (fig. 7.3)—reported at first that Galileo was born on February 18, the exact date of Michelangelo's death, but the 18 was later changed to 19. I cannot say if this document is reliable, since nothing indicates when and by whom it was written and altered. But even some historians, such as G.B.C. Nelli in the eighteenth century, have claimed that Galileo was born on February 18.[26] Gherardini was perhaps the most cautious: He left the date blank (". . . .").[27] Favaro believed that Viviani's date was incorrect, possibly based, as we shall soon see, on a document that had been tampered with.[28]

Galileo's birth was recorded in a number of documents, some of

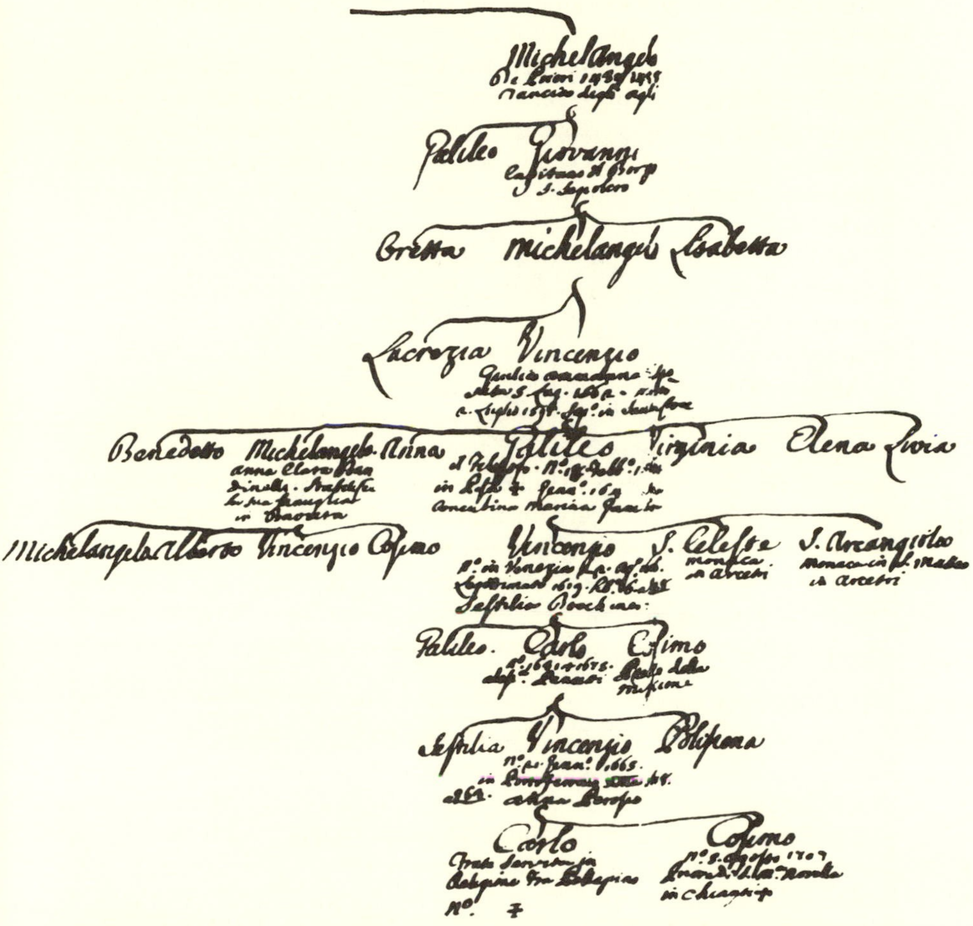

Figure 7.3. Galilean MS 11, folio 161. The date of Galileo's birth has been changed from 18 to 19 February. (Courtesy of the Biblioteca Nazionale Centrale, Florence)

which Viviani may not have seen, including at least five different horoscopic charts. The horoscopes, which can be taken as reliable, at least for birthdays, all agree on February 15.[29] Two were published in the National Edition, of which one (fig. 7.4) actually says February 18, but, as Favaro pointed out, close examination clearly shows that the "8" was originally a "5." Favaro, however, failed to notice that the "5"

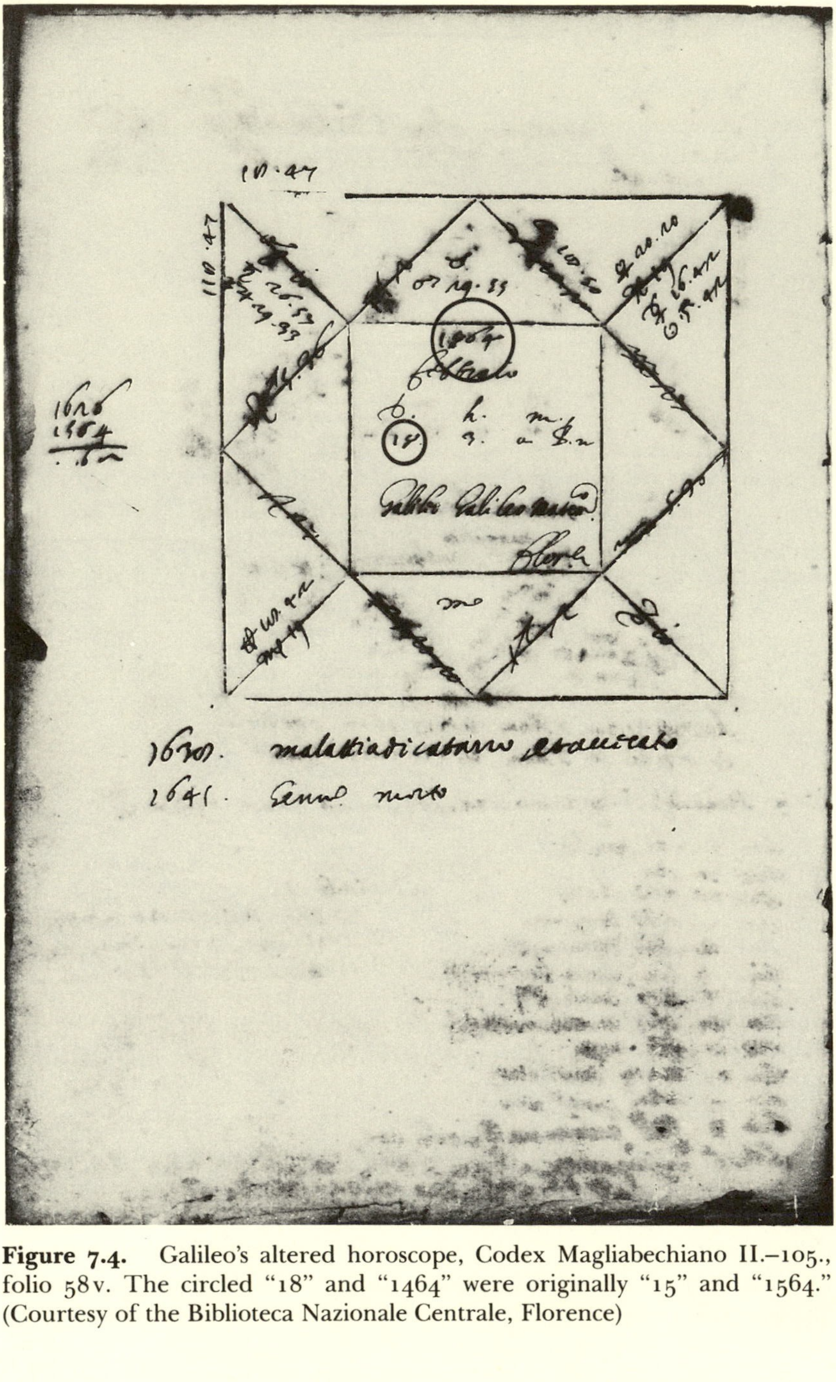

Figure 7.4. Galileo's altered horoscope, Codex Magliabechiano II.–105., folio 58v. The circled "18" and "1464" were originally "15" and "1564." (Courtesy of the Biblioteca Nazionale Centrale, Florence)

in "1564" also had been altered, to "4," turning 1564 into 1464. These two changes are not visible in figure 7.4 because it is black and white; in the original document the brownish revised numbers are distinctly different in color from the rest of the writing. This indicates that the newer numbers were written with a different ink, which may have faded differently than the original. Perhaps the changes were not noticeable when fresh.

Who altered the chart and why? Were both alterations made by the same person? Who would change the year of Galileo's birth to 1464? When was this done? I can say only that the day "15" was probably changed to "18" after the seventeenth century because the "8" differs from the way it was commonly written in Viviani's day (see Viviani's calculations in fig. 7.2), so Viviani can hardly be blamed. Favaro did not explicitly accuse Viviani of falsifying the document, but of having deliberately altered the report of Galileo's date of birth in order to make it coincide with Michelangelo's death. Favaro based his claim chiefly on still unpublished documents in Galilean MS 11 (folios 167–171) containing information related to Michelangelo's death, including the abovementioned excerpt from Vasari. Some of this information is in Viviani's handwriting, in particular folios 168v and 171 (fig. 7.2 and 7.5), which bear strange calculations. According to Favaro, Viviani's calculations were an attempt to "torture chronology" in order to make the day, and even the hour, of Michelangelo's death coincide with Galileo's birth. (Also see figs. 7.2 and 7.6.)

My reading of these calculations, however, does not agree with Favaro. At most, Viviani appears only to have been checking whether the length of Michelangelo's life remained the same in the various dating systems then in use. During the sixteenth and seventeenth centuries, different Italian cities used different dating systems. In Florence, where Viviani was writing, the years were counted *ab incarnatione,* so that they began on the Feast of the Annunciation (March 25). In Pisa, where Galileo was born, the year began on the *preceding* March 25. Thus, the date of Michelangelo's death (in Rome) would have been recorded in Florence as February 18, 1563, but in Rome and in Pisa as February 18, 1564.[30]

What Viviani's calculations do confirm beyond any doubt is that he was following the pattern of classical Renaissance biographies of artists, exactly as described by Kris and Kurz. He urgently sought a link between Galileo and the great Michelangelo and was scrupulously checking the evidence. There is no indication of his trying to distort facts; in fact, his final result, entered in both drafts *A* and *B,* showed Galileo's birth on February 19 (not 18). It seems to me that Nelli, a

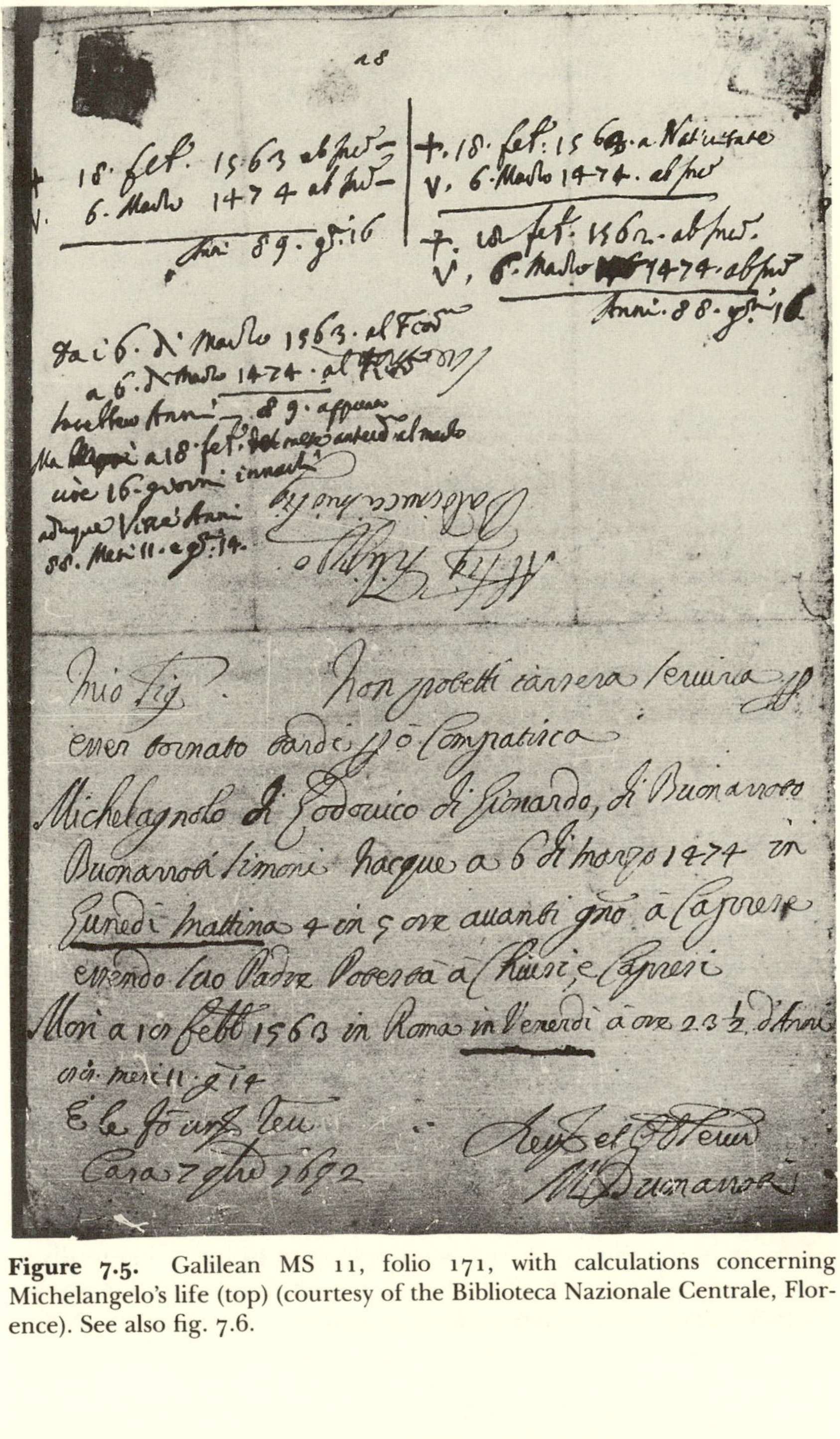
18

+. 18. feb. 1563 ab Inc.
V. 6. Marzo 1474 ab Inc.
Anni 89. g. 16

+. 18. feb. 1563 a Nativitate
V. 6. Marzo 1474. ab Inc.
+. 18 feb. 1562. ab Inc.
V. 6. Marzo 1474. ab Inc.
Anni. 88. g. 16

Dai 6. di Marzo 1563. al Fior.
a 6. di Marzo 1474. al Fior.
Sarebbero Anni 89. appunto
Ma Morì a 18 feb. del mese avanti al Marzo
cioè 16 giorni innanzi
adunque Visse Anni 88. Mesi 11. e g. 14.

Mio Sig. Non potetti iersera servirla per
esser tornato tardi, mi compatisca
Michelagnolo di Lodovico di Lionardo, di Buonarroto
Buonarroti Simoni nacque a 6 di marzo 1474 in
Lunedi mattina 4 o 5 ore avanti giorno a Caprese
essendo suo Padre Podestà a Chiusi, e Caprese
Morì a 18 febb 1563 in Roma in Venerdi a ore 23 ½ d'Anni
88. mesi 11. g. 14
E le fo' umil. reu.
Casa 7 bre 1692
Dev. et Obl. Servitore
M. Buonarroti

Figure 7.5. Galilean MS 11, folio 171, with calculations concerning Michelangelo's life (top) (courtesy of the Biblioteca Nazionale Centrale, Florence). See also fig. 7.6.

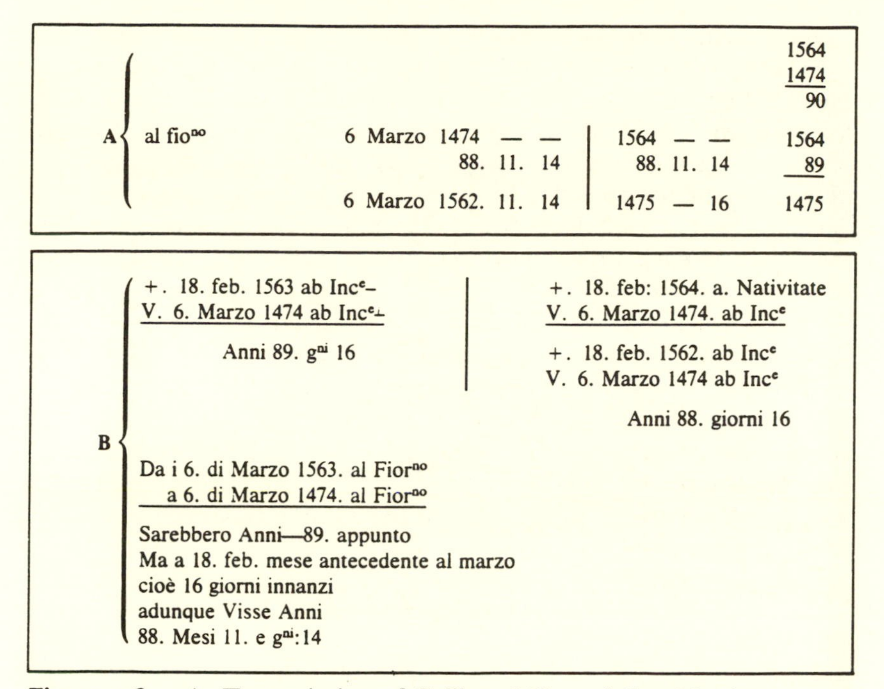

A { al fio^{no}

		1564
		1474
		90
6 Marzo 1474 — —	1564 — —	1564
88. 11. 14	88. 11. 14	89
6 Marzo 1562. 11. 14	1475 — 16	1475

B {

+. 18. feb. 1563 ab Inc^{e}–
V. 6. Marzo 1474 ab Inc^{e}

Anni 89. g^{ni} 16

+. 18. feb: 1564. a. Nativitate
V. 6. Marzo 1474. ab Inc^{e}

+. 18. feb. 1562. ab Inc^{e}
V. 6. Marzo 1474 ab Inc^{e}

Anni 88. giorni 16

Da i 6. di Marzo 1563. al Fior^{no}
a 6. di Marzo 1474. al Fior^{no}

Sarebbero Anni—89. appunto
Ma a 18. feb. mese antecedente al marzo
cioè 16 giorni innanzi
adunque Visse Anni
88. Mesi 11. e g^{ni}:14

Figure 7.6. A: Transcription of Galilean MS 11, folio 168v, bottom (see fig. 7.2); "al fio^{no}" means "in the Florentine style." B: Transcription of Galilean MS 11, folio 171, top (see fig. 7.5). Bottom section reads: "From the 6th March 1563 in the Florentine style / to the 6th March 1474 in the Florentine style / should be exactly 89 years. / But to 18th February, the month preceding March / namely 16 days earlier / thus he lived 88 / years 11 months and 14 days."

century later, had more reason than Viviani to falsify the horoscopic chart. It was Nelli, who with access to most of the Galilean material, probably this very chart, wrote that Galileo was born on February 18. This does not prove that Nelli was responsible for the forgery.

I was, however, unable to determine how Viviani reached the date February 19. His conclusion seems rather bizarre, considering that in 1692 he obtained a copy of Galileo's baptismal certificate dating the baptism to February 19.[31] In Viviani's time, as today, it was unlikely that a child would be baptized on the day he was born. Although there is no special rule about when a newborn should be baptized, in the seventeenth century the customary wait was at least three days after birth; in winter, to avoid endangering the child's life, baptism was postponed as long as possible. Galileo himself baptized his daughter Virginia nine days after her birth, although she was born in summer.[32]

Yet Viviani selected February 19 for his "Racconto istorico" and wrote it in both drafts *A* and *B*. Why? The answer is problematical. He may have relied on Vincenzio Galilei's notes, which he said he obtained as early as 1666. In 1692, when he obtained a copy of Galileo's baptismal certificate, Viviani was already fairly old and might not have gotten around to correcting his manuscript (in which case his drafts in Galilean MS 11 must have been written after 1666, and there must have been an earlier draft written in 1654). Perhaps he was testing another linkage—between Galileo and Copernicus, who was born on February 19, 1473.[33] But this is only speculation, and there is no evidence that Viviani knew Copernicus's birthday.

The custom of linking great men seems gradually to have gone out of fashion in the eighteenth century, and Salvini (in his 1717 printing) corrected Viviani's report of Galileo's birth date. But throughout his brief biography, Viviani did his best to embellish Galileo's image, what Kris and Kurz called "heroization of the artist in biography."

The Heroization of Galileo

Viviani's description of Galileo as a child prodigy—already "in his early age Galileo showed his talent by building, in his free time, alone, instruments and small machines"[34]—was very nearly copied from Vasari's life of Giotto. Vasari said of young Giotto: "mostrando in tutti gli atti ancora *fanciulleschi* una *vivacità* e prontezza d'*ingegno*" (he "showed in all his *boyish* ways *liveliness* and quick *intelligence*"); Viviani wrote that: "ne' prim'anni della sua *fanciullezza* a dar saggio della *vivacità* del suo *ingegno*" ("in the early years of his *youth* he showed the *liveliness* of his *intelligence*").[35]

Viviani added that Galileo was self-taught, read his first Latin authors while still very young; learned Greek, played the lute, and was so talented as a painter that he could have become a professional artist. This may all have been true, but no supporting evidence was given. It all conforms so neatly to the general pattern of a Renaissance biography that it would not be surprising if Viviani had invented it, or at least embroidered the facts.

After recounting Galileo's childhood, Viviani described his university studies. Here the variations between drafts become especially interesting: They considerably stretch the truth and often reinforce doubts raised on other grounds. He variously gave Galileo's age on entering the university as eighteen (in *A* and *B*), seventeen (in *B*), and, finally, sixteen (also in *B*).[36] According to documents in the archives of the studium of Pisa, Galileo began his studies in 1581, when he was over seventeen-and-a-half.[37] Similarly, Viviani varied the age at which he said Galileo started studying mathematics, with Ostilio Ricci. First he said it was twenty-two (in *A* and *B*), then reduced it to nineteen (in *B*).[38]

Ricci's role in Galileo's life was like Cimabue's alleged role in Giotto's life, and Gherardini included an anecdote about Galileo, typical of a Renaissance biography, that echoed the supposed casual meeting between Giotto and Cimabue. According to this story, Galileo acquired his first mathematical notions by hiding outside the room in which Ricci was tutoring the pages of the grand duke.[39]

Viviani added other curious details about Galileo's initiation into mathematics: In manuscript *A* he said Galileo had learned geometry in "pochi mesi" (a few months), then in *B* crossed out the word "mesi" with a pencil and substituted "tempo" (time). Viviani clearly intended to write "a short time," which is vague and could sound even shorter then "a few months," but neglected to alter the declension of *pochi* to *poco*, resulting in the grammatically incorrect "pochi tempo."[40]

One possible instance of this kind of image enhancement may actually have been a response to new information. In manuscript *A* Viviani wrote that at the age of twenty-four Galileo had already formulated the Appendix to his *Two New Sciences*, dealing with the centers of gravity of solid bodies; in *B* he lowered the age to twenty-one. In fact, Galileo did write this treatise in his youth, but it is not known exactly when.[41] Still, Viviani's revised figure may very well have been based on what he considered a more accurate finding.

Viviani also remarked, just as classical biographers of artists did, that Galileo had a particularly deep knowledge of nature: "Nature chose Galileo as the one who should reveal a part of those secrets."[42] Galileo's alleged discovery of the principle of the pendulum at the age of nineteen after observing a lamp swinging in the Cathedral of Pisa was a typical illustration of the fathoming of a whole principle from a single particular. The pendulum in Pisa may be considered analogous to the proverbial lion's claw.

Viviani's and Gherardini's story about the Swedish king, Gustavus Adolphus, does appear to have been apocryphal, since there is no evidence that the king was ever in Italy; also, he was only sixteen years old in 1610, when Galileo left Padua. But two Swedish princes, both called Gustavus or Gustav, apparently did spend some time in Italy and may even have studied with Galileo, and one was the son of the deposed King Erik XIV.[43] Thus, this attempt to link Galileo with a royal personage, like many of Viviani's and Gherardini's stories, had a kernel of truth adorned with imaginary details.

The most famous and controversial anecdote in Vivian's "Racconto istorico" was, of course, the story of the Leaning Tower experiment. In theory it should hold little importance for either science or science history because, even if true, it certainly had no impact on Galileo's thinking. If it did occur, it was only a public performance, and Galileo would not have climbed to the top of the tower without knowing the result beforehand. The aggrandizement of the experiment, as Lane Cooper showed some fifty years ago, was not due to Viviani but was an artifact of the literature that appeared afterward, mainly the popular literature which repeated and amplified Viviani's story to the level of a major event in science history and a triumph of empirical over a priori science. Viviani was more balanced and in his telling of the story also said that Galileo refuted many Aristotelian principles of motion "by means of experiments and of solid demonstrations and arguments."[44]

How true was the story?[45] The experiment, as was shown in chapter 1, was performed, according to Viviani, when Galileo was a professor in Pisa, more or less in the period when he is said to have written his *De motu.* Since the dating of the various parts of this work is uncertain, it may be risky to relate them to the Leaning Tower experiment. Nevertheless, in *De motu* Galileo repeatedly cited the example of a body falling from a tower, although without describing a specific experiment.[46] Perhaps he did drop things off the tower to test his theory; it is not unlikely that the story had some truth in it. Moreover, during the seventeenth century it was fashionable for professors in Pisa to perform experiments from the top of the Leaning Tower. For instance, Giorgio Coresio, a professor of Greek, reported having done so in 1612,[47] and Vincenzio Renieri reported having performed such experiments repeatedly in 1641.[48] And Carlo Rinaldini, a professor of philosophy who collaborated with Viviani and whom we shall meet again in the next chapter, reported having tried, with a Torricelli tube, to measure the difference in air pressure between the top and bottom of the Tower.[49] If Galileo did not actually do the experiment, Viviani needed little imagination to tell an anecdote that was sufficiently realistic while conforming to the scheme of the biography he was writing.

Viviani made an interesting revision in drafts of his "Racconto" related to the Leaning Tower experiment. In *A* he ended the story by saying "all this is treated extensively [*diffusamente*] by him [Galileo] in his last Dialogues Concerning two New Sciences." In *B* this became first "all this is treated extensively by him in the *said* Dialogues Concerning New Sciences" and then was entirely canceled.[50] Viviani probably thought at first that Galileo's refutation of Aristotle's law of fall appeared in the last Days of the *Two New Sciences,* which dealt with motion. He later realized that Galileo did it in the *First Day* of the "said" work, which dealt mainly with strength of materials, and so wrote "said Dialogues" instead of "last Dialogue." However, Galileo there had challenged Aristotle's law by means of a *thought,* rather than a material, experiment.[51] But from our discussion of the likely character of Viviani's readers we recall that they did not like this sort of nonempirical argument; for them, the Leaning Tower story was much more convincing. This may conceivably have been the reason that Viviani deleted his entire reference to Galileo's statement.[52]

Many other details in Viviani's "Racconto istorico" show that he was conforming to a particular biographical model in vogue at that time. In general, many of Viviani's draft revisions cannot be traced to available documents. The only quoted dates we can now document with any precision are Galileo's birth and the age at which he entered the university, and in both cases Viviani was wrong. It seems that he took advantage of occasions in Galileo's life that could not be authenticated and shaped them to magnify Galileo's aptitude and enhance his empiricist image. However, although Viviani was undoubtedly writing in the tradition of Vasari, and Vasari had invented facts, Viviani (like Baldinucci) was more "scholarly" than Vasari, and his stories had at least a kernel of truth. My reading of Viviani's paper shows that he was generally accurate and tried to document his assertions as well as he could. He was more "modern" as a historian than one might think. His obsessive revisions, in places where he may have invented facts, indicate that altering reality must have been very difficult for him; he was too much of a perfectionist—in our sense of the word—for this type of literature. This aspect of Viviani's character and work may encourage modern historians to give more credit to some of his anecdotes and to understand them better, even when undocumented, always bearing in mind that he felt he had to shade his information to satisfy his readers. Similarly, his emphasis on Galileo's empiricism conformed to the expectation of his audience. Viviani may have intended such emphasis to be slight, but it was expanded out of proportion by later biographers.

Placing Viviani's "Racconto istorico" in its historical context does not give a final answer to the question of Galileo's empiricism, but it shows that a consideration of contemporary reports in their correct historiographical context can help to avoid misinterpretation. As far as Galileo's image as an empiricist is concerned, it suggests, at least, that we may have been asking the wrong question. Instead of asking which experiments described by Viviani, or by Galileo himself, did Galileo actually perform, perhaps we should ask why did Viviani (and Galileo) think it important to report these experiments? This might be a better approach to evaluating, in general, early accounts of the work of Galileo and his followers and, in particular, their most "empiricist" enterprise: the Accademia del Cimento.

8

The Accademia del Cimento

The Triumph of Empiricism

We are now nearing the end of our story. In the previous chapters I tried to show how complex and problematic the work of Galileo and his followers was, involving, for instance, philosophical, ideological, theological, social, political, and even personal and psychological factors. The contribution of Galileo and his school can certainly not be characterized as merely laying the groundwork for experimental science. On the contrary, there is little evidence that experiment was the key element of their work. Neither Galileo nor his major followers—Cavalieri and Torricelli in particular—appear to have attached particular importance to experiment. Rather, Galileo's empiricist image was a later creation inspired by Viviani's life of Galileo, among other sources.

Yet reports of scientific activity at the Tuscan court in Viviani's day seem to contradict this picture and to confirm the empiricist image of Galilean science. Was this science actually experimental, or do the reports reflect an accepted view of science, dictated by external factors?

Before attempting to answer the question, let me outline the scientific activity in Tuscany in Viviani's time. Despite the prevailing decadence, Grand Duke Ferdinand II de' Medici, an amateur scientist, "un principe filosofo," as Targioni Tozzetti called him, continued to promote the letters and sciences in Tuscany, for instance, by granting fellowships. Beneficiaries included Viviani, Carlo Dati, and Francesco Redi (one of the leading contemporary biologists in Italy, also a man of letters and a poet, who served the court as chief physician).[1] When assigning the fellowship to Viviani, the grand duke told him:

> I have granted you the salary as that of a lecturer in mathematics, not so that you lecture; Redi and Dati do not lecture: these are honorary lectureships, which we grant to those who are good in writing. Let me

> know when you have something ready to publish; I shall see that somebody else lectures in Pisa. They [the lecturers] will lecture to a small audience and you will write for everybody, in present and future. You write things, real things, whereas they will say words which are carried away by the wind.[2]

Ferdinand's patronage was not limited to financial support; he promoted and also took an active part in several projects. The projects included, according to Targioni Tozzetti, the experimental incubation of eggs, performed in 1664 in the Boboli garden (garden of the Tuscan court in Florence) by experts invited from Cairo. Ferdinand was also said to have invented several instruments, including, in 1649, a new type of thermometer containing spirit instead of water, thus lowering the range of temperature measurement below the freezing point of water. The thermometer was used in various experiments, from meteorological studies to measurement of the freezing point of liquids to measurement of the temperature of brooding hens.[3]

Targioni Tozzetti also found some documents that, he believed, described the work of a scientific academy active at court from 1653, though the evidence is too slim to be conclusive.[4] In 1657, however, scientific activity at the Tuscan court became established and regular under Ferdinand's hardworking brother, Leopold, and was recorded in diaries that began on June 19, 1657. This was how the Accademia del Cimento came into being.

The Accademia soon became famous and was later regarded as the peer of the French Académie Royale des Sciences and the English Royal Society of London. It was active only until 1667—why it stopped then is still not clear—and then published the results of its work in the *Saggi di naturali esperienze* (here abbreviated to *Saggi*) written, under the supervision of Prince Leopold, by its young secretary, Lorenzo Magalotti. But they were also reported in dozens of volumes of unpublished manuscripts now in the Galilean collection in Florence.[5] These include diaries, descriptions of experiments, observations of various sorts, correspondence, and other accounts of scientific activity. Two relatively broad studies of the Accademia have been made. One, mentioned in chapter 2, was Targioni Tozzetti's in the eighteenth century; the second is a modern one by W. E. K. Middleton. Both rely on documents still mostly unpublished, but we are still far from knowing the whole truth.

The Accademia del Cimento is generally considered a direct heritage from Galileo, a sort of scientific monument erected by Tuscany's rulers in Galileo's memory, and in fact has much in common with Galileo's work. It investigated the fields of physics and astronomy that had also been Galileo's province, and its leading members, Borelli and Viviani, were Galileo's leading followers (I found no evidence that Borelli ever met Galileo personally, but he had been part of Castelli's "Galilean School").

The impression made by the work of the Accademia del Cimento is that it was unequivocally empirical. Its name, *Cimento,* to begin with, means experiment (the 1680 edition of the *Vocabolario della Crusca* did not define cimento but referred the reader to the terms "esperimentare" and "esperimento"). It may also be related to the goldsmith's *cimentare,* meaning to assay (gold), which also implies "putting to the test."[6] The Accademia's motto, "provando e riprovando," taken from a passage in Dante's *Divine Comedy* and displayed on its coat of arms, also indicated empiricist tendencies; it may be translated as "testing and retesting," or "testing and refuting."[7] Finally, we have the *Saggi di naturali esperienze* (*Essays on Experiments in Nature; esperienza* in modern Italian means experience, but in the seventeenth century also meant experiment), which contained descriptions of experiments. Its introduction deplored the limited returns attainable with geometry in favor of experiment as a more immediately productive scientific tool:

> [Geometry] leads us a long way along the road of philosophical speculation, but then abandons us when we least expect it. This is not because geometry does not cover infinite spaces and traverse all the universal works of nature in the sense that they all obey the mathematical laws by which the eternal understanding governs and directs them; but because we ourselves have not as yet taken more than a few strides on this long and spacious road. Now here, where we are no longer permitted to step forward, there is nothing better to turn to than our faith in experiment.[8]

Although the preference for experiment is evident throughout the *Saggi,* was the Accademia del Cimento as empirical as it appeared? Let us consider in more detail its institutional setting and some aspects of its activities.

A Strange Scientific Society

The Accademia del Cimento, entirely devoted as it was to the investigation of nature, differed from most of its contemporaries, at least in Italy, which were largely literary academies. Yet even as a scientific academy it was peculiar.

First, unlike other scientific academies of that time, it had no official statutes or regulations. The Lincean Academy, founded more than sixty years earlier, already had a comprehensive declaration of principles regulating the conduct of its members down to such minute details as the length of meetings. Moreover, no document so far found, if one ever existed, shows that the Accademia del Cimento was ever officially and publicly registered. We do not know how the Accademia officially admitted its members, what their privileges and duties were, or even how they met. Its records, still unpublished, indicate only that its members met in a private and informal way, in different places (usually depending on the whereabouts of its patron, the prince), and that activities were interrupted for long periods. We do not even know for sure the names of all the members and their collaborators. Some are mentioned in the records but, as Targioni Tozzetti noted, there may have been others.

All historians agree that the best mind and the driving spirit of the Accademia was Borelli. Borelli was approaching the age of fifty, had behind him a long and successful scientific and academic career in Sicily, had published works in mathematics and physiology, and was the leading scientist in Italy. He received the chair of mathematics in Pisa in 1656, but despite the prominence of its previous incumbents Galileo and Castelli, the chair was not particularly prestigious, and Borelli may have deserved more. At least, he apparently expected better treatment and expressed disappointment when the overseer of the university, Filippo Magalotti, informed him that he was obliged to teach.[9] Borelli may have wanted to be treated like Galileo and Viviani, who had not had to teach and at most had been expected to serve the court. Nevertheless, he seems from the beginning to have expected to play a dominant role in the Accademia del Cimento, since his arrival in Tuscany coincided with its first "official" activity at the Tuscan court.

Of course, Borelli was influenced by Galileo, and one would have expected him, as the leading member of the Accademia, to be assisted by other Galilean followers. Surprisingly, however, Viviani was the only other member who was a declared and leading follower of Galileo. Two others who may have been linked to Galileo, the brothers Candido and Paolo del Buono, were obscure intellectuals. Candido, of whom little is known, was a priest who may have learned mechanics directly from Galileo. His younger brother Paolo never took part in academy meetings and contributed only through correspondence; in the ten years the Accademia was active, Paolo was abroad, mostly in Vienna, in the service of the Holy Roman Emperor.

Most of the other members had little to do with physical research in general and with Galilean science in particular. They were physicians,

Figure 8.1. The Accademia del Cimento. (Engraving by G. Vascellini in *Serie di ritratti d'uomini illustri Toscani con gli elogi storici dei medesimi,* vol. 4, Florence 1773)

Aristotelian philosophers, and men of letters. Antonio Oliva (or Uliva), a professor of medicine at the University of Pisa, did assist Borelli in several experiments carried out in the Accademia.[10] Francesco Redi also was probably a member.[11]

The other members, Alessandro Marsili and Carlo Rinaldini, were Aristotelian professors of philosophy at the University of Pisa, and Rinaldini was also Prince Leopold's Philosopher, possibly one of the reasons he was accepted into the Accademia. Carlo Dati, too, may have been a member, or at least suggested a number of experiments.

How did this widely diverse group of intellectuals gather? How did they collaborate, if they did at all? Before answering these questions, let us consider an enigmatic aspect of the Accademia del Cimento: its official publication, the *Saggi.*

The *Saggi* purported to summarize ten years of work, including 253 groups of experiments, but was a rather restricted sample compared to the forty-nine volumes of the Accademia's manuscripts. Most of the experiments were in physics (the last group studied digestion in animals), largely related to the behavior of matter and its various states. The *Saggi* contained almost no theory—at most some references to philosophers such as Plato, Plutarch, Gassendi, Galileo, Boyle, and a few others. It did not indicate by whom and when the experiments were carried out and offered almost no comments on results. In many cases it records observations with no apparent preconceived view and without leading to or even attempting to draw conclusions.

The *Saggi* seem to have been published first and foremost to impress the reader with form rather than content. Even for the Baroque Period, decorations were excessive and represented an enormous amount of work for the publication. Targioni Tozzetti commented, not without scorn, that "the woodcut initials, the decorations and the tailpiece are too large, suitable for a choir book and unsuited to the text of the page."[12] Henry Oldenburg, secretary of the Royal Society of London, on receiving a copy of the *Saggi* in 1668, remarked in a letter to Robert Boyle that the book was "pompous" (in the seventeenth century "pompous" usually meant "magnificent").[13] Magalotti, in fact, had had the Royal Society's copy of the *Saggi* specially and sumptuously bound. Yet Oldenburg added that it contained nothing new except for a few experiments on amber.

The *Saggi,* as well as other unusual aspects of the Accademia del Cimento, can be partially explained by the Accademia's total dependence on its patron, well documented by Middleton. It was de facto Prince Leopold's private academy, and as such had no need for statutes or regulations; the prince decided everything. He must have chosen whom to include in the Accademia (he must have been extremely idealistic, or at least optimistic, to believe that such a heterogeneous group of intellectuals would collaborate constructively), where and when it met, what it should investigate, and which results should be published. Although the prince was learned in science, he remained primarily a politician—in practice, the most powerful man in Tuscany, more influential than the grand duke himself—and often made his decisions on political, rather than purely scientific, grounds.

Thus, to understand the work of the Accademia del Cimento, it must be considered—much as in Galileo's case—on two different levels: the scientific and the political. The scientific level constitutes its true scientific work (most of it unpublished); the political level has to do with its public image, which was an expression of the will of the prince, represented by formal aspects such as its name, its motto, and the *Saggi.* How, if at all, do these aspects relate to each other?

The Accademia's unpublished manuscripts dealing with experimental physics, like the *Saggi,* contain mainly descriptions of empirical works.[14] Meteorological observations (no fewer than twelve volumes of manuscripts), in particular, contain only numerical records.[15] However, the manuscripts do show that members embraced and discussed various physical and astronomical theories that went far beyond the *Saggi* reports. To illustrate this, let us consider the Accademia's astronomical investigations and its experiments and discussions on atomism.

The Astronomy of the Accademia

Abetti and Pagnini, the modern editors of the *Saggi* (1942), noted that the astronomical activity of the Accademia del Cimento, recorded in four volumes of manuscripts, was only marginal and was restricted to particular topics, as if designed more to give the appearance of keeping pace with the progress of astronomy in Europe than to execute a preconceived program.[16] Indeed, the *Saggi* mentioned none of this work. Yet it was in astronomy that the Accademia del Cimento made one of its most important contributions to seventeenth-century science—confirmation of Huygens's theory that Saturn was surrounded by a ring.

Aimed at explaining the strange, cuplike, and varying appearance of Saturn, this was also the only work of the Accademia that was eventually published and studied satisfactorily.[17] The mystery of Saturn had begun in 1610 when Galileo discovered its unusual appearance and conjectured that the planet was accompanied by two "stars" on its sides, like Jupiter with its four satellites. But these "stars" of Saturn behaved unlike Jupiter's satellites and for nearly fifty years remained unexplained. In 1659, Christiaan Huygens suggested, in his *Systema Saturnium,* that Saturn was surrounded by a ring. His unorthodox theory contradicted not only Aristotle's view that planets are spherical and incorruptible but also Galileo's explanation.

Huygens's theory was soon contested in Rome by Honoré Fabri, a Jesuit, mathematician, Inquisitor, and corresponding member of the Accademia del Cimento. Fabri suggested, in a compromise between

Aristotle and Galileo, that the "handles" of Saturn were not a ring but an optical effect produced by a combination of five satellites revolving behind the planet. He accepted Galileo's opinion that Saturn, like Jupiter, had small bodies near it and even increased their number. Yet he argued that these small "planets" did not rotate around Saturn but about points behind it. His explanation was thus closer to Aristotle's view that all celestial bodies rotate around the earth. Fabri's theory was not published directly by him but in a work by Eustachio Divini, one of the best telescope makers of the time.[18]

Both Huygens and Divini dedicated their publications to Prince Leopold, a proof of the Medici's successful preservation of the family fame as patrons of the new science. Furthermore, Divini ended his work with an appeal to Prince Leopold to arbitrate between the two theories, and Huygens wrote a reply to Fabri in the form of a letter addressed to the prince. Leopold and his academy were being asked to arbitrate between a Protestant Dutchman (Huygens) and a member of the Inquisition (Fabri), a delicate but prestigious task that indicated the high reputation of the Accademia del Cimento in its early days. The task was rendered even more delicate by the fact that Fabri was officially a member of the Accademia. But the Accademia, as its heterogeneous membership suggests, had hardly any collegial solidarity.

To evaluate the two theories, Academicians performed a series of tests. In one, both interesting and original, they made a model of Huygens's theory and invited unbiased simple people (*persone idiote*) to observe the model from a distance and report what they saw. The test was not as immune to preconceptions as it may appear; had the Academicians been genuinely free of prejudice, they would have repeated the experiment with Fabri models. But other tests, including the first observation of the shadow cast by Saturn's main body on its rings, confirmed that Huygens's hypothesis was more likely to be true. The final verdict was in favor of Huygens, and even Fabri finally accepted the ring explanation.

Saturn was not the only astronomical body the Accademia del Cimento studied. During the winter of 1664–1665 it made observations on a comet, primarily to refute claims that comets were sublunary phenomena. Here the Academicians were challenging not only Aristotle but also Galileo, both of whom had claimed that comets were merely optical effects. They were again successful. As Borelli suggested to Prince Leopold, two or more observations of the comet at different locations showing a very small parallax between the two lines of sight would suffice to disprove the Aristotelian claim.[19] The idea was not new; Grassi, the Jesuit mathematician against whom Galileo had argued, had made the same point more than forty years earlier.[20] Leopold obtained records of observations made in Paris by the French astronomer Adrien Auzout, and documents relating to this research contain a table comparing observations from Rome, Mantua, Pisa, and Florence.[21] Trigonometrical calculations proved that the comet was a celestial event.

Among other objects of astronomical observations by the Accademia were the solar eclipse of 1661, the moons of Jupiter, the lunar eclipse of 1666, and Venus. The *Saggi* reported none of these, and some were published for the first time only a century later, by Targioni Tozzetti.

Atoms Again

The Accademia del Cimento engaged in other theoretical discussions more directly related to the *Saggi,* but well disguised by its experimental descriptions. One in particular involved atoms. Many of the experiments included in the summary at the end of the *Saggi* related to heat, changes in state of substances, and topics such as pneumatics, natural and artificial freezing, and thermal expansion.[22] Unpublished Accademia correspondence shows that members not only advanced theories to explain empirical findings but often pictured matter as composed of tiny particles. The question of atomism seems to have been much debated, and one source of inspiration was Pierre Gassendi, the contemporary French atomistic philosopher (as Middleton remarked, Gassendi received four mentions, nearly as many as Galileo's six).[23] The range of the Accademia's deliberations remains to be studied, but the correspondence shows their strong interest. One illustrative instance concerned the nature of heat.

In a letter (1658) to Prince Leopold, Borelli conjectured, just as Galileo had thirty-five years earlier in *The Assayer,* that heat was nothing but a substance composed of fiery particles, that heated bodies increased their volume by absorbing these particles, and that they lost volume by ejecting them.[24] It seems that other Academicians disagreed, evidently because Borelli's theory did not explain the irregular behavior of water around its freezing point. They postulated two types of particle, hot and cold. A note by Magalotti, discovered by W. E. K. Middleton and probably addressed to Viviani, said "E viva gl'atomi frigorifici" (long live the atoms of cold).[25] Borelli countered with the argument that

> The volume of some fluids, e.g., *aqua vitae* and quicksilver, at any degree of cooling, not only does not increase, but even decreases. And, no doubt, had the expansion of freezing water depended on the intrusion of those refrigerated bodies, the quicksilver and the *aqua vitae* should necessarily have dilated by the increasing effort of cold. *Vice versa,* assuming that cooling is nothing but a diminution of igneous bodies, a greater cold should necessarily cause a decrease in volume and this suits the experiment, while the other [case] does not. Therefore, cold is deprivation of heat matter [*caldella*].[26]

Borelli suggested two experiments to settle the matter, which exemplify the type of experiment the *Saggi* recorded without mention of preliminary discussions. In this particular case, the *Saggi* allotted only about half a sentence to the theoretical background—: "to throw some light on the question, whether the cooling of a body results from the entry of some kind of special atoms of cold"—followed by a short description of an experiment and a carefully phrased conclusion: "Instead of an excess of matter coming out, it seemed rather to demonstrate an evacuation or loss of something (if it was not really the contraction of that which was there), much water being sucked to fill the space."[27]

Ensuing debate centered around whether heat was a substance or a quality. Middleton studied the manuscripts and reported that the Academicians performed an experiment in September 1657 that was recorded in one of the two (still unpublished) diaries of the Accademia.[28] The experimenters placed thermometers above and beside a hot ball and tried to explain why the upper thermometers registered higher temperatures than the lower ones, hoping to resolve the question about the nature of heat. The interpretations varied with the views of the interpreter. Rinaldini, and probably also Marsili, took the Aristotelian view that heat is a quality. Rinaldini argued that "The quality of heat, warming the ambient air, makes the hotter parts ascend, and the cooler remain below; so that the warmer parts surrounding the upper thermometer make it move more that the lower one, which is surrounded by the parts of air that are less warm."[29] The opposing view, undoubtedly held by Borelli, was that heat is not a quality "because if it were, it would appear that it should diffuse equally in every direction, exactly as is asserted by the Peripatetics. But if it consists of corpuscles [i.e., of air], as claimed by Democritus, it does not seem difficult to understand that they would move mainly upward."[30]

Although further experiments yielded no conclusive results, these examples demonstrate that the Accademia designed experiments to a large extent to test existing theories, such as Aristotle's or Galileo's, or even new theories hypothesized by the Academicians. As Middleton made clear, the Academicians were not merely Baconian "collectors of facts"; they also discussed theories ranging from world systems to the atomic structure of matter.

Indeed, evidence shows that not all the Academicians agreed with the empirical methodology implied by the *Saggi.* Targioni Tozzetti, for instance, published a curious allegorical essay by Carlo Dati, entitled "Utility and Delight of Geometry," that argued against an empirical scientific method. In an eloquent and figurative style, Dati described the vicissitudes the human soul goes through in its search for truth. Having tried philosophy in vain, the soul turns to experiment, but realizes that this action is not based on sound principles. Sobbing in despair, the soul is finally approached by a maiden who announces, "Your troubles are over—I, geometry, am the remedy."[31] Dati's essay does not prove that Italian scientists of the time extensively debated the status of experiment, as did Boyle and Hobbes in England during the same period.[32] In Italy the debate seems to have been more muted (to be heightened by later writers). But it does indicate that empiricism was not the only tendency in the Accademia del Cimento.

Why, then, did the empirical work of the Accademia del Cimento figure so prominently in the *Saggi?* The Accademia del Cimento, like the Accademia della Crusca, may have been trying to fulfill general contemporary expectations. However, there is another explanation, one that attributes the responsibility to the Church's interference in the activities of Galileo's followers.

The Church and Galileo's Followers

According to the generally accepted view, the Church intimidated Galileo's followers and prevented them from carrying on the work of their teacher, even in fields like physics and mathematics that were considered less "heretical" than astronomy. The Church's general policy might be described as what Lanfranco Belloni calls "Bellarminism": discouragement of any attempt to mathematicize physics.[33] A "neutral" empirical practice of science was therefore much safer. Thus, instead of risking controversy with the Church, Prince Leopold may have chosen to conceal some of the Accademia's deliberations, especially in astronomy, and had published work that appeared prudently empirical. This painful choice would prevent Galileo's followers from receiving some of the credit they deserved and withhold from the Medici some of the prestige they sought, but would also avoid a clash with the Church and give the Accademia del Cimento freer hand sub rosa. Such a decision by the prince can help us to understand why the *Saggi* were so selective; committing work on astronomy and other controversial topics to writing would have been asking for trouble.

An effort to avoid conflict with the Church might, in theory, also explain the heterogeneous membership of the Accademia. Leopold may have been trying to disguise, or at least to balance, the Accademia's Galilean approach by enrolling scientists with different inclinations. Marsili and Rinaldini, the two Aristotelian philosophers, appear to have formed a pole opposite that of the Galilean Borelli. This may not have been premeditated, but the Accademia del Cimento was, in a way, structured like Galileo's *Dialogue,* in which one of the contending interlocutors, Simplicio, was an Aristotelian philosopher. All in all, had it not been for this policy, Galileo's disciples might have been persecuted. The prince, who ran the academy, can thus be regarded as an enlightened supporter of Galilean science who not only helped foster it but was also able to preserve it.

All this may have been true; yet the interaction between the Church and Galileo's followers was much more complex than is generally assumed. One must remember that most of Galileo's followers were ecclesiastics. Castelli was a Benedictine, Cavalieri a Jesuat; Vincenzio Renieri an Olivetan friar. Ricci became a Cardinal; Magiotti, Nardi, and Berti (about whom we know little), if not also ecclesiastics certainly served or acted in Roman ecclesiastical circles; and Torricelli and Viviani had been trained by the Jesuits. Hence it was not the Church but some factions within it that may have interfered in the work of Galileo's followers. Galileo's main opponents were either Jesuits or members of the Inquisition orders, Dominicans and Franciscans. Since very little research has so far been conducted on the archives of these orders, evidence is too meager to indicate how much they actually influenced the work of the Accademia del Cimento.[34]

For instance, in a 1665 letter to Prince Leopold de' Medici, Borelli proposed the publication in France of observations he had made on the comet that appeared at the beginning of that year "so that people on this side of the mountains may see the free way of speaking [in France] in assemblies of Jesuits and other men of letters, and how everybody speaks of the Pythagorean [i.e., Copernican] system, so that the sentence [of Galileo in 1633] may become acceptable and less frightening." Although this letter indicates that some Italian scientists were afraid in the 1660s, it also implies that in Borelli's view there was no real danger.[35]

In another example, given by Targioni Tozzetti, Viviani "worked in secrecy and being rightly afraid of some search, kept all the writings of Galileo, of his followers and of his correspondents in a pit in his house."[36] But Favaro commented that fear may not necessarily have been Viviani's motive. Targioni Tozzetti, in his great admiration for Galileo's struggle for Copernicanism, may have exaggerated a little.[37]

Perhaps somewhat sounder evidence comes from a letter written in 1658 by Carlo Rinaldini to Prince Leopold warning that in view of the forthcoming publication of the *Saggi* "the Jesuits are making a din before time; they say that if that book of natural observations contains anything concerning some of them, they will have the appropriate people to give an answer." However, was this a threat of the Inquisition against "forbidden science," or were the Jesuits just a competing school promising to air a contradicting opinion? Paolo Galluzzi, who published part of Rinaldini's letter in 1981, suggests that since Rinaldini was basically anti-Galilean, his letter may have been an attempt to hinder the work of the Accademia.[38]

Other examples suggesting Church intimidation may exist, but my failure to find more than these few tends to indicate that the danger was less than is generally assumed. I do not claim that ecclesiastical suppression played no prime role, but simply point out that it was the Medici, rather than the Church, who directly controlled the scientists in the Accademia del Cimento, and that such restraint was sometimes overdone. The "self-censorship" imposed on the Accademia del Cimento by Prince Leopold de' Medici may not really have been necessary.

Borelli's published writings indicate that Leopold's caution was largely excessive. Nothing in fact happened to Borelli when he published two works on the astronomical observations he made during his stay in Tuscany, which were part and parcel of the work of the Accademia del Cimento. One, on the comet of 1664–1665 and published under a pseudonym, was based on observations he made together with other Academicians. The other purported to be a physical account of the motion of Jupiter's moons but was actually a disguised (and outstanding) study of the Copernican system.[39] It seems evident, therefore, that by paying a little attention to the manner of presentation, the Accademia could have allowed much more to be published than it did. In fact, a note (probably written by Lorenzo Magalotti) to

Prince Leopold showed that it even planned to publish astronomical material on Saturn, provided that "Borelli, to avoid difficulty, content himself with keeping his proofs outside the Copernican system."[40] However, this material was not published. Why?

I suggest that the main concern of the Medici was not specifically to prevent a clash with the Inquisition, but rather to maintain an eclectic attitude toward knowledge so that they could present themselves as the patrons of one "universal" scientific culture. This would be, in a way, in line with Galileo's own attitude, that is, stressing judiciousness while satisfying all the institutions and individual intellectuals concerned with science (whether Galilean or not), including, of course, the Inquisition.[41] As Borelli said on one occasion: "These princes try to avoid a clamorous appearance that might arouse malevolence and clamor, and in short [they see] that true philosophy spreads in a pleasant way and soft manner."[42] Experimental physics (as presented by the *Saggi*) served this purpose in the best way. It did not involve theoretical discussions or controversies and could satisfy both Galilean and non-Galilean scientists. Regrettably, forcing different scientific approaches into one framework was not appropriate for an academy like the Accademia del Cimento that intended to carry out a specific program.

Whatever the reason for the Accademia del Cimento's emphasis on empirical aspects, it seems to have had more to do with Medici patronage than science itself. And this late Medici patronage, as I will argue in the following Epilogue, contributed to the decline, rather than the advancement of Galilean science.

Epilogue: The Decline of Italian Science

I end with some short general remarks on the decline of Italian science during the second half of the seventeenth century. After the Accademia del Cimento ceased to be active, in 1667, little scientific activity was recorded in Tuscany, or in the rest of Italy.

According to the traditional view, the Church was the main cause of the decline of Italian science. This view is by no means false; the Church certainly hindered the work of Galileo and his followers, and evidence indicates a growing disfavor toward Galileanism all over Italy toward the end of the century. For instance, in 1671 Honoré Fabri (himself an Inquisitor) was tried by the Roman Inquisition and sentenced to fifty days' imprisonment for suggesting that proof of the earth's motion, if found, could be accommodated by a more symbolic interpretation of relevant passages in the Bible.[1] Later, between 1688 and 1697, a few members of the Academy of Investigators and some atomists in Naples were tried as atheists.[2] In 1691 Grand Duke Cosimo III, under Jesuit pressure, officially forbade the study of atomism. His edict reads:

> No professor of the University of Pisa may read or teach, either publicly or privately, whether in writing or by speech, the Democritan philosophy, namely about atoms, but only the Aristotelian, and whoever should in any way transgress the wish of the Grand Duke may, in addition to the indignation of His Highness, consider himself summarily dismissed.[3]

And in 1693 Antonio Baldigiana, Consultor of the Holy Office, wrote from Rome to Viviani:

> There have been held, and are being held extraordinary congregations of cardinals of the Holy Office and before the pope, and there is talk of a general prohibition of all authors of modern physics; long lists of them are being made, and these are headed by Galileo, Gassendi and Descartes as most pernicious to the literary republic and the sincerity of religion. The chief persons to form a judgment of them will be the religious, who at other times have made efforts to issue these prohibitions.[4]

These instances indicate a new wave of repression all over the peninsula, toward the end of the century, that inhibited the remaining supporters of Galileo. Yet the fact remains that, as far as Tuscany before 1667 is concerned, Galileo's followers could carry on their master's work relatively undisturbed. The accepted belief that the Church paralyzed Italian science has diverted attention away from the work of Galileo's followers, as if it were insignificant. But—as I have tried to show in this book—their scientific activities, at least during the quarter of a century after his death, were broad, complex and far from negligible. And their study suggests the decline had additional causes, both social and scientific. We have seen that the Galilean revolution left in its wake far from trivial scientific problems. Galileo's followers proposed various solutions, not always satisfactory. They often disagreed among themselves and at times were faced with controversies and scandals. Of course, pressure also came from ecclesiastical circles, but naming the Church as the sole cause of the decline of Italian science is somewhat of an exaggeration.

A decisive factor in the decline of Galilean science, in particular, may well have been the scarcity of Galileans to continue Galileo's work. Galileo seems to have been so busy with politics that he neglected for too long the task of training a new generation of scientists. Castelli and his school were instrumental in disseminating Galilean ideas, and Galileo would probably have had many other active followers if he had nurtured more pupils like Castelli. Most of the few remaining active Galileans—Castelli, Cavalieri, Torricelli, Nardi, and Vincenzio Renieri—died shortly after Galileo, in the 1640s, and Magiotti died in 1658. This left only Viviani, Borelli, and perhaps Ricci (in Rome), and only Borelli was a first-rate scientist worthy of taking Galileo's and Torricelli's place. Galilean science soon had virtually no active supporters. It might have faded away even without pressure from the Church.

One other factor in the decline, primarily affecting Tuscany during a particular period, may have been the suffocating patronage of the Medici. After Galileo's death, his successors continued for a time without interference by the Medici, which permitted Torricelli, for example, to make outstanding contributions to seventeenth-century science. But after Torricelli's death, the Medici began to exert more control over research in the physical sciences. It is true that the court's support was vital for scientists like Borelli and for the establishment of the Accademia del Cimento. But it is also true that Prince Leopold de' Medici unnecessarily forced a restraining policy on his scientists, imposed censorship, and in general created a climate that hindered their work. As a result, the Accademia del Cimento was dissolved, research in the physical sciences at the Tuscan court stopped in less than a decade, and Galileo's followers were given much less credit than they deserved for their achievements; even today much of their scientific work is still unpublished.

Notes

Introduction

1. Galileo was born on February 15, 1564. His date of birth is sometimes given as February 18 or 19 to make it coincide with Michelangelo's death in Rome on February 18, 1564 (I will discuss this linking in chapter 7). Shakespeare was born on April 26, 1564 and Newton on January 4, 1643, by the Gregorian calendar. Many authors have remarked that Newton was born the year Galileo died; this is true only if Newton's birth is dated December 25, 1642 according to the Julian calendar, which was then still used in England. Pope Gregory XIII reformed the Julian calendar in 1582, but England did not adopt the New Style until 1752.

2. Admittedly, no documentary evidence supports this explanation of Galileo's trial and condemnation (see M. A. Finocchiaro, *The Galileo Affair,* pp. 11–13). Nevertheless, Finocchiaro, too, noted that Galileo was, after all, a Catholic living in a Catholic country during a period when decisions of the Council were taken seriously and implemented.

3. There are, of course, many books on the history of Florence. A useful history is J. R. Hale, *Florence and the Medici.*

4. Tuscany was one of Urban's major targets; he persistently urged the Tuscan clergy to act against the grand duke. In 1627, before assuming direct rule of his country, Ferdinand II de' Medici visited Rome and was ill-treated by some members of the Barberini family. During the plague of 1630, the Church forced him to cancel some hygienic precautions he had proclaimed. In 1631, Urbino, a small duchy between Tuscany and the Papal States claimed by both as part of their territory, was annexed to the Papal States. Finally, in 1642, Tuscany joined Edoardo Farnese, duke of Parma, in his campaign to reconquer the city of Castro from the State of the Church. The Pope was defeated, and his humiliation may have hastened his death in 1644.

5. The plague in Florence and its economic implications have been studied by Carlo Maria Cipolla in *Cristofano and the Plague: A Study in the History of Public Health in the Age of Galileo* (London: Collins, 1973); *I pidocchi e il Granduca* (Bologna: Il Mulino, 1980); and *Fighting the Plague in Seventeenth Century Italy* (Madison: Univ. of Wisconsin Press, 1980).

6. Universities during the Middle Ages are studied in Hastings Rashdal, *The Universities of Europe in the Middle Ages,* 3 vols. (Oxford: Clarendon Press, 1895; London: Oxford Univ. Press, 1936). The state of knowledge during the Middle Ages is described in David C. Lindberg (ed.), *Science in the Middle Ages* (Chicago and London: Univ. of Chicago Press, 1978). This collection includes (pp. 120–144) an article by Pearl Kibre and Nancy Siraisi, "The Institutional Setting: The Universities," dealing mainly with the universities of Bologna, Padua, Paris, and Oxford.

7. Indeed, a broad literature deals with single academies but relatively little with academies in general. A short article on the Italian academies in Galileo's day, focusing on the Florentine Academy, is Eric Cochrane, "Le Accademie," in G. Garfagnini (ed.), *Firenze e la Toscana dei Medici nell'Europa del' 500* 1 : 3–17. To the best of my knowledge no general study has been made of the Italian academies up to the sixteenth century. Most Italian academies are listed by Michele Maylender, *Storia delle accademie d'Italia,* 5 vols. (Bologna: Cappelli, 1926–1930). This is an encyclopedic work, not a general study of academies, whose great variety may perhaps make such a study impossible. On the early Italian academies see Frances A. Yates, *The French Academies of the Sixteenth Century* (London: The Warburg Institute, 1947), chap. 1. Also, a useful collection of articles on Italian academies is Laetitia Boehm and Ezio Raimondi (eds.), *Università, accademie e società scientifiche in Italia e in Germania dal cinquecento al settecento* (Bologna: Il Mulino, 1981).

8. A Greek manuscript, which reached Florence in 1460, contained most of the Corpus Hermeticum, a collection of treatises of magic practices written during the second century but attributed to Hermes Trismegistus, a legendary figure believed to have lived in Moses' time. The discovery was considered exceptional, and Cosimo asked Ficino to interrupt this translation of Plato and to translate the Corpus into Latin. Ficino did so in the years 1463–1464, and his translation gave an impetus to the practice of cabala, magic, and occultism in general. Many editions and translations into other languages followed throughout the fifteenth and sixteenth centuries.

9. The special attention and encouragement the Medici gave to cabalistic and non-Aristotelian philosophy are evident in a letter written in 1638 to Grand Duke Ferdinand II de' Medici by Tommaso Campanella (1568–1639), the heretic philosopher and cabalist who had spent most of his life in prison (Fabroni, *Lettere,* 1 : 1). Campanella says among other things: "I, and any men of distinguished talent, were very indebted to the Medici princes, who, by bringing to Italy the books of Plato which were not seen by our ancients, freed us from the yoke of Aristotle and all his followers; and Italy began to learn the philosophy of nations through reason and the experience of nature, and not only the words of men."

10. Various articles in Garfagnini (ed.), *Firenze e la Toscana,* cover many aspects of the cultural policy of the Medici. The importance Cosimo I attached to culture as a major state project is apparent in a letter of 1587 from Antonio Lupicini, a hydraulic engineer in the service of the Tuscan court, to Grand Duke Ferdinand I, published by Iodoco del Badia, "Egnazio Danti, cosmografo e matematico," *Rassegna nazionale* (1881) 4 : 27–28.

11. The main source for the history of the studium of Pisa in Galileo's time is Angelo Fabroni, *Historia academiae pisanae . . .* , 3 vols. (Pisa, 1791–1795). For a modern study see Giovanni Cascio Pratilli *L'università e il principe* (Florence: Olschki, 1975) and Charles B. Schmitt, "The Faculty of Arts at Pisa at the Time of Galileo," *Physis* (1972) 14 : 243–272, reprinted in *Studies in Renaissance Philosophy and Science* and in "The Studio Pisano in the European Cultural Context of the Sixteenth Century," in Garfagnini (ed.), *Firenze e la*

Toscana, vol. 1, pp. 19–36. Cosimo's reform of the University of Pisa is outlined by Danilo Marrara, *L'università di Pisa come università statale nel granducato mediceo* (Milan: A. Giuffrè, 1965).

12. In general, Cosimo attempted to reduce the control of the ecclesiastical authorities over higher education, for example, ordering convents in Pisa having their own studia to close them. Nevertheless, he acted with a certain diplomacy and did not hesitate, in 1564, to enforce a bull of Pope Pius IV obliging students wishing to graduate to take an oath to the ecclesiastical authorities in addition to an oath to the state. The groups subject to this mandatory oath included non-Catholic students, mainly Protestants from abroad and Jews. In spite of many appeals for leniency, the Medici remained uncompromising on this issue until the eighteenth century. Their intransigence was partly dictated by a determination to avoid non-Catholic (i.e., Protestant) influence and to preserve the country from religious conflicts.

13. Cosimo's new statutes also prohibited Tuscan subjects from receiving degrees from private institutions, and, after 1551, even from studying or teaching abroad without permission of the authorities.

14. One of those invited was Vesalius, who in 1544 delivered a number of lectures in Pisa.

15. Cochrane, "Le accademie."

16. Ibid. For a history of the University of Florence see Giovanni Prezziner, *Storia del pubblico studio e delle società scientifiche e letterarie di Firenze* (Florence, 1810).

17. "Filosofia è verace conoscimento delle cose naturali, e delle divine, e delle umane, tanto quanto l'huomo è possente d'intendere."

18. For the various interpretations of and approaches to Aristotle during the Renaissance, see Charles B. Schmitt, *Aristotle and the Renaissance* (Cambridge, Mass., and London: Harvard University Press, 1983).

19. W. E. Knowles Middleton defines philosophy and mathematics in "Science in Rome, 1675–1700, and the Accademia Fisicomatematica of Giovanni Giustino Ciampini," *BJHS* (1975) 8:138–154, p. 144. Middleton quotes this definition from a letter dated July 23, 1677, sent by the secretary of this Roman academy to Appollinare Rocca in Reggio Emilia. A history of the Italian renaissance of mathematics is presented in Paul Lawrence Rose, *The Italian Renaissance of Mathematics.*

20. On the abacus schools see Enrico Gamba and Vico Montebelli, "La matematica abachistica tra ricupero della tradizione e rinnovamento scientifico," in Manno (ed.), *Cultura, scienze e tecniche,* pp. 169–202.

21. Stillman Drake, in "Tartaglia's Squadra and Galileo's Compasso," *AIMSSF* (1977) anno 2, fascicolo 1:35–54, claims that Galileo's geometrical and military compass was, at least, derived partly from Tartaglia's elevation gauge for guns.

22. *Nicholas Copernicus Complete Works,* vol. 2, *On the Revolutions,* edited by Jerzy Dobrzycki, translated into English with comments by Edward Rosen (Warsaw and Cracow: Polish Academy of Science, 1978), p. XVI. Italics are mine.

23. On the complexity of the methodological (as well as juridical and epistemological) background of Galileo's trial, see Maurice A. Finocchiaro, "The Methodological Background to Galileo's Trial," in William A. Wallace (ed.), *Reinterpreting Galileo* (Washington D.C.: The Catholic University of America Press, 1986), pp. 241–272. A thorough treatment of the epistemological background of Galileo's trial appears in Morpurgo-Tagliabue, *I processi di Galileo e l'epistemologia.*

24. I quote from the subtitle of Owen Gingerich, "The Galileo Affair," *Scientific American* (August 1982) 247: 119–127. See p. 3 (Table of Contents).

25. "Astronomia. Scienzia che tratta del corso de Cieli e delle stelle. But la qual cosa, che intenda ora chiaramente dimostra per ragione astrologica." I have never come across the word "but" in Italian, but this is exactly what the *Vocabolario* says. Also, the *Vocabolario* itself has no such entry in its collection of words.

26. Alistair C. Crombie in "Galileo in Renaissance Europe," Garfagnini, *Firenze e la Toscana,* 2: 751–762. A. C. Crombie says that Galileo was a *virtuoso* (in the Italian sense), which in Italy at the end of the sixteenth century meant a man with a wide, nonspecialized culture and with the intellectual power to command any situation. But Galileo was also more than a virtuoso; he had both an impressive general, especially literary culture, and very specialized mathematical and philosophical knowledge.

Chapter 1. Galileo: The Public Figure

1. Thomas B. Settle, "Ostilio Ricci, a Bridge between Alberti and Galileo," *Actes, XIIe Congrès International d'Histoire des Sciences, Paris, 1968,* Tome III-B: *Science et Philosophie, XVIIe et XVIIIe Siècles* (Paris: Librairie Scientifique et Technique Albert Blanchard, 1971), pp. 121–26.

2. *OG* 19:633–646, pp. 636–637.

3. "Racconto istorico . . . ," *OG* 19:597–632, p. 603.

4. Galileo's early works in manuscript have been collected in *OG* 1.

5. Collected in Gal. MSS 27 and 46 respectively. Favaro did not publish the first set (except for a summary in *OG* 9: 281–282 and an excerpt in *OG* 9: 290–291) because he regarded it as an insignificant collection of notes from the period in which Galileo studied in Vallombrosa. The second set was published in *OG* 1: 15–177, under the title *Juvenilia.* These manuscripts have lately been subject to several studies. According to William A. Wallace, *Prelude to Galileo: Essays on Medieval and Sixteenth-Century Sources of Galileo's Thought* (Dordrecht, Boston, London: Reidel, 1981) and *Galileo and His Sources: The Heritage of the Collegio Romano in Galileo's Science* (Princeton: Princeton Univ. Press, 1984), they seem to date around 1590, and therefore the first set could not have been written in Vallombrosa. See also William A. Wallace, "Reinterpreting Galileo on the Basis of His Latin Manuscripts," in W. A. Wallace (ed.), *Reinterpreting Galileo,* pp. 3–28. Pages 3–5 summarize the story of these manuscripts. A. Carugo and A. C. Crombie date *Juvenilia* much later, after

1597. See "The Jesuits and Galileo's Ideas of Science and of Nature," p. 60. Winifred L. Wisan, "On the Chronology of Galileo's Writings," *AIMSSF* (1984) anno 9, fascicolo 2:85–88, discusses this chronology.

6. Gal. MS 71, collected in *OG* 1:245–419. Translated into English with Introductions and Notes by I. E. Drabkin in Galileo, *On Motion and On Mechanics,* pp. 3–131. Carugo and Crombie (ibid.) date this work to much later.

7. *OG* 19:606.

8. Galileo's letter to Mazzoni is more of a short scientific treatise refuting Mazzoni's argument against Copernicanism (in his book *In universam Platonis et Aristotelis philosophiam praeludia sive de comparatione Platonis et Aristotelis* (Venice, 1597) than a letter, and was therefore published in *OG* 2:193–202 separately from Galileo's correspondence. The letter to Kepler is in *OG* 10:67–68. The evolution of Galileo's Copernicanism is examined by Maurice A. Finocchiaro in "Galileo's Copernicanism and the Acceptability of Guiding Assumptions," in *Scrutinizing Science: Empirical Studies of Scientific Change,* edited by Arthur Donovan, Larry Laudan, and Rachel Laudan (Dordrecht: Kluwer Academic, 1988), pp. 49–67. In considering this (pretelescope) stage of Galileo's Copernicanism (pp. 53–57), Finocchiaro concludes that Galileo's cognitive stance may be described as a partial pursuit only.

9. Favaro, *Galileo Galilei e lo studio di Padova,* 1:61–62.

10. *OG* 10:83–84.

11. *OG* 10:115–116.

12. Koyré, *Galileo Studies,* part 2, pp. 63–123, "The Law of Falling Bodies: Descartes and Galileo." Galileo's wrong inference is also discussed by Marshall Clagett, *The Science of Mechanics in the Middle Ages* (Madison: University of Wisconsin Press; London: Oxford University Press, 1959), p. 580, and Stillman Drake, "The Uniform Motion Equivalent to a Uniformly Accelerated Motion from Rest (Galileo's Gleanings XX), *Isis* (1972) 63:28–38. A critical analysis of the passage in Galileo's *Two New Sciences* related to this problem is made by Maurice A. Finocchiaro in "Vires Acquirit Eundo: The Passage Where Galileo Renounces Space-Acceleration and Causal Investigation," *Physis* (1972) 14:125–145, and "Galileo's Space-Proportionality Argument: A Role for Logic in Historiography," *Physis* (1973) 15:65–72.

13. This prohibition is often disregarded in the literature of the history of science, but not in the many biographies of Sarpi. See, for instance, Federico Chabod, *La politica di Paolo Sarpi* (Reprint. Venice-Rome: Istituto per la Collaborazione Culturale, 1968), chapter 2. Galileo himself describes the expulsion of the Jesuits in a humorous letter written in 1606 to his brother, Michelangelo, *OG* 10:158.

14. The various stages of the invention of the telescope are described by Albert Van Helden in "The Invention of the Telescope," *Transactions of the American Philosophical Society* (1977) vol. 67, part 4.

15. The translation of the title is explained by Drake, *Discoveries and Opinions,* p. 19.

16. *OG* 10:233; translation from Drake, ibid., 65. Gaetano Cozzi, in comparing the patronage choice by Galileo with that by Benedetti a few decades

earlier, draws attention to the fact that both ultimately preferred the patronage of absolute rulers to that of the Venetian republic. See "La politica culturale della Repubblica di Venezia nell'età di Giovan Battista Benedetti," in Manno (ed.), *Cultura, scienze e tecniche,* pp. 9–27.

17. The importance of Galileo's political campaign is also emphasized by Richard S. Westfall in "Science and Patronage: Galileo and the Telescope," *Isis* (1985) 76:11–30.

18. *OG* 10:348–353.

19. *OG* 10:369, 372–375, 400–401.

20. See Biagioli, "Galileo the Emblem maker."

21. Drake, *Galileo at Work,* pp. 166–167. A brief history of the Lincean Academy is given by Drake in "The Accademia dei Lincei," *Science* (1966) 151:1194–1200.

22. Geymonat, *Galileo Galilei,* p. 58 of the English translation.

23. Koestler, *The Sleepwalkers.* This book is sometimes unjustly regarded as dilettantish, perhaps because it claims that Galileo's arrogance was the main cause of his misfortunes.

24. Agassi, *Towards an Historiography of Science,* pp. 56–57.

25. *OG* 5:7–260 and partly translated into English with an introduction and notes by Stillman Drake under the title "Letters on Sunspots" in *Discoveries and Opinions,* pp. 59–85.

26. Not in March, as Santillana mistakenly reports in his *The Crime of Galileo* (p. 39). I point out this mistake because the reports of Galileo's trial, including Santillana's, indicate that things could change radically within a very short time. According to Santillana, the debate took place on March 13. But Castelli reports in a letter dated December 14, 1613 (a Saturday): "Thursday morning I was at the table . . ." (*OG* 11:605); hence the debate most probably took place on December 12. Santillana was probably confused by the fact that the month of March had, more than once, been crucial for Galileo. In March 1613 Galileo published his controversial *Letters on Sunspots,* and in March 1616 Copernicus's *De revolutionibus* was suspended. Stillman Drake, in his *Discoveries and Opinions,* p. 151, also mistakes the day of the debate and says it occurred on a Wednesday; the mistake is corrected in *Galileo at Work,* p. 222.

27. "Lettera a D. Benedetto Castelli," December 21, 1613, *OG* 5:279–288; translated into English by Maurice Finocchiaro in *The Galileo Affair,* pp. 49–54.

28. *The Sleepwalkers,* p. 433.

29. The copy reached Bacon through Toby Matthews. See *OG* 12:255.

30. *OG* 19:298.

31. *Lettera a Madama Cristina di Lorena,* 1615, *OG* 5:307–348; translated into English with an introduction and notes by Stillman Drake in *Discoveries and Opinions,* pp. 145–216, and by Maurice Finocchiaro in *The Galileo Affair,* pp. 87–118.

32. *OG* 5:327: "Bisogna, prima che condannare una proposizion naturale, mostrar ch'ella non sia dimostrata necessariamente: e questo devon fare non

quelli che la tengon per vera, ma quelli che la stiman falsa." English translation from Finocchiaro, *The Galileo Affair,* p. 102. This well-known statement has sometimes been produced out of context, indicating the priority of conclusively established scientific knowledge over biblical statement. See Maurice A. Finocchiaro, "The Methodological Background to Galileo's Trial," in W. A. Wallace *Reinterpreting Galileo* pp. 241–272, especially p. 268.

33. *OG* 12:171–172. A translation of the letter appears in Finocchiaro, *The Galileo Affair,* pp. 67–69. Foscarini's book was entitled *Lettera . . . sopra l'opinione de' Pittagorici e del Copernico della mobilità della terra e stabilità del sole* (Naples, 1615).

34. Drake, *Galileo at Work,* p. 256.

35. The controversy, with translations of the various related essays, is presented in Stillman Drake and C. D. O'Malley (trans.), *The Controversy on the Comets of 1618* (Philadelphia: University of Pennsylvania Press, 1960).

36. Redondi, *Galileo Heretic.*

37. *OG* 19:321–322; translation in Finocchiaro, *The Galileo Affair,* pp. 147–148.

38. It is also not entirely clear whether Galileo was tortured or not. Although a great majority of historians say it was very unlikely, it has recently been reasserted that torture was part of the usual procedure in such a case, even if the accused was as old as Galileo. See Italo Mereu, *Storia dell'intolleranza in Europa* (Milan: Mondadori, 1979), pp. 386–414. Morpurgo-Tagliabue, in an addendum to the 1981 edition of his *I processi de Galileo e l'epistemologia* (pp. 198–199), argues against this view.

39. The quotations are from Finocchiaro, *The Galileo Affair,* p. 291. Finocchiaro (p. 363, n. 86) says that the technical term "vehementemente sospetto d'heresia" (*OG* 19:405) seemed to indicate a specific category of crime, second in seriousness only to heresy.

40. *OG* 19:402–407; Finocchiaro, *The Galileo Affair,* pp. 287–291. The quotation is from Santillana, *The Crime of Galileo,* p. 293.

41. As Agassi makes clear in his *Towards an Historiography of Science.*

Chapter 2. Galileo's Image through the Ages

1. A. Koestler, *The Sleepwalkers;* A. Koyré, *Galileo Studies* and "An Experiment in Measurement," *Proceedings of the American Philosophical Society* (1953) 97:222–237. Republished in *Metaphysics and Measurement,* pp. 89–117.

2. Galileo's image as a martyr of science is treated by Maurice A. Finocchiaro in "The Methodological Background to Galileo's Trial," in William A. Wallace (ed.), *Reinterpreting Galileo,* pp. 241–272, and in "Toward a Philosophical Reinterpretation of the Galileo Affair," *Nuncius* (1986) anno 1, fascicolo 1:189–202.

3. The prince intended to include the biography in Galileo's *Opere* then being edited in Bologna by Carlo Manolessi. The whole story is related in a 1656 letter by Viviani, *Carteggio* 2:301–308.

4. Viviani's biography, the "Racconto istorico" (*OG* 19:597–632), was pub-

lished for the first time by Salvino Salvini in *Fasti consolari dell'Accademia Fiorentina* (Florence, 1717), pp. 397–431. Gherardini's biography (*OG* 19: 633–646) was published for the first time by Targioni Tozzetti, *TT* 2:62–76.

5. *OG* 19:625; italics in the English translation are mine: "Dicendo che i caratteri con cui era scritto erano le proposizioni, figure e conclusioni geometriche, per il cui solo mezzo potevasi penetrare alcuno degli infiniti misterii dell'istessa natura. Era perciò provvisto di pochissimi libri, ma questi de' migliori e di prima classe: lodava ben sì il vedere quanto in filosofia e geometria era stato scritto di buono, per delucidare e svegliar la mente a simili e più alte speculazioni; ma ben diceva che le principali porte per introdursi nel ricchissimo erario della natural filosofia erano l'osservazioni e l'esperienze, che, per mezzo delle chiavi de' sensi, da i più nobili e curiosi intelletti si potevano aprire." For the first part of the translation I relied on A. C. Crombie, "Galileo in Renaissance Europe," in Garfagnini (ed.), *Firenze e la Toscana,* 2:751–762, especially p. 751. For Crombie, in Viviani's essay, "The intellect was in command" (p. 752), showing how differently Viviani's description can be interpreted. In the present chapter I shall refer, on purpose, to Galileo's "empiricist" rather than "empirical" image, where "empiricist" means empirical to an improper degree.

6. *OG* 19:603.

7. *OG* 19:606. Italics are mine.

8. *OG* 19:638: "S'accopiarono in lui lo speculare e l'operare, la teorica e la pratica."

9. Galileo's bibliography was first compiled by A. Carli and A. Favaro in 1896. It was updated in 1943 by G. Boffito, and in 1967 (up to 1964) by Ernan McMullin, as a supplement to his *Galileo: Man of Science.*

10. *TT.* Targioni Tozzetti was a physician with a wide range of interests—botany, mineralogy, paleontology, and the history of science. He traveled around Tuscany, observed and described its nature, and collected plants and fossils. He was able to contribute to the history of science thanks to his work as a librarian in the Florentine Magliabechian library, which he reorganized (some of the catalogues he made are still being used). Copies he made of important scientific documents benefit modern historians of science. He also began writing a monumental history of science in Tuscany from the time of the Etruscans to his day, which was interrupted by his death and remained mostly unpublished. The manuscript, entitled *Selve* (seventeen volumes), is now in the National Library of Florence (MSS Palatini, Serie Targioni Tozzetti, 189).

11. Nelli, *Vita e commercio letterario di Galileo Galilei.*

12. The story of Nelli's discovery of Galileo's manuscript is related in *TT* 1:124–125. Favaro, throughout his *Galileo Galilei e lo studio di Padova,* criticized Nelli's work as inexact and misleading.

13. The documentation used by Nelli and the drafts of his work are collected in MSS 318–322 of the Galilean Collection.

14. Alexandre de Humboldt, *Cosmos: Essai d'une description physique du monde,* trans. H. Faye, 4 vols. (Paris 1846), 1:189; Favaro, *Galileo Galilei e lo studio di Padova,* 1:13, n. 2.

15. Antonio Favaro, "Sul giorno della nascita di Galilei."

16. Wohlwill first expressed his doubts at a meeting reported in the *Münchener medizinische Wochenschrift* (1903) 50:1849–1850 and later, in detail, in his series of articles "Galilei-Studien": "Die pisaner Fallversuche," *Mitteilungen zur Geschichte der Medizin und der Naturwissenschaften* (1905) 4:229–248; "Der Abschied von Pisa," *Mitteilungen zur Geschichte der Medizin und der Naturwissenschaften* (1906) 5:230–249, 439–464, (see pp. 230–231 on the pendulum); (1907) 6:231–242. Favaro argued against Wohlwill in "Sulla veridicità del 'Racconto istorico della Vita di Galileo' dettato da Vincenzio Viviani," and "Di alcune inesattezze nel 'Racconto istorico della Vita di Galileo' dettato da Vincenzio Viviani." The discussion is summarized by Giacomelli in *Galileo Galilei giovane e il suo "De motu"*, pp. 1–23. On Wohlwill's life and work, see Hans-Werner Schütt, *Emil Wohlwill: Galilei-Forscher, Chemiker, Hamburger Bürger im 19. Jahrhundert* (Hildesheim: Gerstenberg, 1972).

17. Galileo's letter of 1602 to Guidobaldo del Monte concerning the pendulum may be found in *OG* 10:97–110.

18. Part of *De motu* was published for the first time under the title *Sermones de motu gravium,* in *Le opere di Galileo Galilei,* edited by Eugenio Alberi, (1854) 11:1–80. In his *De motu,* Galileo says in one of his contradictions to Viviani: "For it is true that wood moves more swiftly than lead in the beginning of its motion; but a little later the motion of the lead is so accelerated that *it leaves the wood behind it.* And if they are both let fall from a high tower, the lead moves far out in front. This is something I have often tested." (*OG* 1:334, translation from Drabkin, p. 107; italics are mine). Yet we still do not know the exact period in which Galileo compiled his *De motu;* see I. E. Drabkin, "A Note on Galileo's *De motu,*" *Isis* (1960) 51:271–277. On the genuineness of the Leaning Tower experiment, see my "Galileo, Viviani and the Tower of Pisa."

19. Cooper gives a long list of authors who added imaginary details to Viviani's story in *Aristotle, Galileo, and the Tower of Pisa.*

20. Ornstein, *The Role of Scientific Societies* (1963), p. 218.

21. Olschki, *Galilei und seine Zeit.* One chapter of this book, "Galileis literarische Bildung," has been translated into English by Thomas Green and Maria Charlesworth, under the title "Galileo's Literary Formation," in McMullin (ed.), *Galileo: Man of Science,* pp. 140–159. Quotation from p. 141.

22. Arnold Pacey, for instance, in *The Maze of Ingenuity, Ideas and Idealism in the Development of Technology* (London: Lane, 1974), believes that men like Galileo, from the Middle Ages to the present day, contributed much more to technology than vice-versa.

23. John Herman Randall, Jr., "The Development of Scientific Method in the School of Padua," *Journal of History of Ideas* (1940) 1:177–206; A. C. Crombie, *Robert Grosseteste,* chapter 11. The influence of the Paduan environment, in general, on Galileo had already been emphasized at the end of the last century by Antonio Favaro in *Galileo Galilei e lo studio di Padova.*

24. Generally speaking, Zabarella distinguished between two processes of discovery: "resolutive," by which knowledge is induced from observation, and

"analytical," by which the cause or "principle" of the effect is defined. According to Zabarella discovery proceeded through four stages: First, he observed a particular effect; second, he resolved the complex fact into its component parts and conditions; third, he examined this supposed, or hypothetical, cause by "mental consideration," to clarify it and its essential elements; and finally, he demonstrated the causes.

25. See Paolo Rossi, "The Aristotelians and the 'Moderns': Hypothesis and Nature," *AIMSSF* (1982) anno 7, fascicolo 1:3–28. Rossi criticizes the alleged influence of Zabarella on Galileo as part of the more general continuity theory in the history of science.

26. Charles B. Schmitt in "Experience and Experiment: A Comparison of Zabarella's View with Galileo's *De motu*," *Studies in the Renaissance* (1969) 16:80–138. Reprinted in *Studies in Renaissance Philosophy and Science.*

27. Thomas B. Settle, "Galileo's Use of Experiment as Tool of Investigation," in McMullin (ed.), *Galileo: Man of Science,* pp. 315–337. See also Settle, "An Experiment in the History of Science."

28. Drake has written many articles on this subject, summarized in his *Galileo at Work: His Scientific Biography* (see p. 415 on the Leaning Tower experiment) and in "Galileo's Notes on Motion," Supplemento agli *AIMSSF* (1979) fascicolo 2, monografia n. 3. For a recent new interpretation of some of these manuscripts, see David K. Hill, "Dissecting Trajectories: Galileo's Early Experiments on Projectile Motion and the Law of Fall," *Isis* (1988) 79:646–668.

29. I have summarized the discussion of Galileo's experimentalism in my "The Role of Experiment in Galileo's Physics."

30. On the contradiction between Descartes's a prioristic methodology and his empirical activity, see A. C. Crombie's article in *DSB* 4:51–55. Spinoza, too, held that sense perception cannot offer a true description of the essence of things but nevertheless performed experiments, making it clear that he regarded empirical work as merely an aid to theoretical work, which, by itself, cannot prove the issue.

31. The importance of studying theories in their historical context has been emphasized by Thomas Kuhn in *The Structure of Scientific Revolutions* and by Joseph Agassi in *Towards an Historiography of Science.*

32. Attention to the criticism of Galileo in his own day has been drawn by Paolo Galluzzi in "Vecchie e nuove prospettive torricelliane," G. Arrighi et al., *La scuola galileiana,* 13–51, pp. 34–46.

33. *OG* 7:47–53. In Drake's translation, pp. 23–28.

34. *Harmonie Universelle,* 1, livre second, 108–112; p. 112: "Je doute que le sieur Galilee ayt fait les experiences de cheutes sur le plan, puis qu'il n'en parle nullement, et que le proportion qu'il donne contredit souvent l'experience."

35. Settle, "An Experiment in the History of Science," p. 21.

36. *OG* 8:212–213. Crew and de Salvio, pp. 178–179.

37. Obviously, the constant of acceleration of the earth had not yet been discovered in Mersenne's and Galileo's time, and Mersenne formulated his

claim in a totally different way. I have "translated" it into modern formulae for the purpose of pinpointing his mistake.

38. Alexandre Koyré, "An Experiment in Measurement," pp. 226–229.

39. P. Ioannis Baptista Ricciolo, *Almagestum Novum Astronomiam veteram . . .* (Bologna, 1651). I rely on Koyré's report of Riccioli's experiments in "An Experiment in Measurement," pp. 229–232. Galileo proclaimed the isochronism of the pendulum and the relation between its length and period in his *Two New Sciences, OG* 8:139–141. Crew and de Salvio, pp. 95–98. For the repetition of the experiment see Ronald Naylor. "Galileo: Real Experiment and Didactic Demonstration," *Isis* (1976) 7:398–419, p. 401.

40. See James MacLachlan, "Galileo's Experiment with Pendulums: Real and Imaginary," *Annals of Science* (1976) 33:173–185. This useful article presents the passages in Galileo's *Dialogue* and *Two New Sciences* in which the various aspects of pendular motion are discussed. On p. 183 it shows, among other things, that Galileo's critics, including Mersenne, are not always in conflict with what Galileo said.

41. Gal. MS 130.

42. Gal. MS 130, 972: "Nel'esperienza della palla sdruciolante per un canale si reputa da me sicura, oltre che il moto di essa è composto di due, mentre scendendo si ruzzola per il sostegno, e per l'aderenza al canale in se stessa: e si come per l'aderenza, e sostegno quella riesce più tarda d'un'altra che per l'aria discenda."

43. Gal. MS 130, 972: "La nuova scienza del Galilei intorno al moto de'i cadenti, e de' i proietti, s'appoggia tutta a due principii, l'uno che il moto orizontale sia uguale, l'altro che il moto de'i cadenti riceva nuove aggiunte di velocità secondo la ragione de'i tempi. Dubito nondimeno che questo secondo principio non bene con l'esperienze concordi, siche non tanto si velociti un cadente quanto da esso principio segue."

44. Standards of acknowledgement are discussed by Joseph Agassi in "Who Discovered Boyle's Law?," *Studies in History and Philosophy of Science* (1977) 8:189–250. He says, for instance, (p. 193): "It is alleged that Kepler privately complained that in Galileo's Sidereal Messenger an acknowledgment to Bruno and to Kepler himself in missing. Now we know enough of Kepler's generosity to doubt that he could so complain; nevertheless, it is quite likely that he did think Galileo had missed an occasion to express indebtedness. This may be unpleasent, but it is not a case of violation of accepted standards: accepted standards, inasmuch as one could at all discuss them, were hopeless. Galileo even made his own standards as he went along: he was engaged in a controversy about priority concerning the discovery of the sunspots and in the course of the debate he stated that priority must go to the one who adequately explained the discovery phenomenon, namely to Galileo himself."

45. This period was also characterized by anticlerical feeling, and the fact that Caverni was a priest may also have worked against him.

46. The story of Caverni's work is related by Giorgio Tabaronni, at the beginning of the reprinted work, and by Cesare S. Maffioli, in "Sulla genesi e sugli inediti della *Storia del metodo sperimentale in Italia di Raffaello Caverni,*"

AIMSSF (1985) anno 10, fascicolo 1 : 23–85. Maffioli's research relies on Caverni's and Favaro's correspondence, among other sources. The 1985 reprint did not include all of Caverni's work, only that contained in the 1900 edition. Volume 6 abruptly (and incredibly) ends in the middle of a sentence.

47. Guillaume Libri, in his *Histoire des sciences mathématiques en Italie, depuis la Renaissance des lettres jusqu'a la fin du dix-septième siècle,* 4 vols. (Paris, 1838–1841), is perhaps the only historian before Caverni to give an account of the contributions to science from ancient times to Galileo. Yet Libri says explicitly (4 : 159) that Galileo had no predecessors.

48. Duhem, *Études sur Léonard de Vinci* and *Le système du monde.* Duhem's approach, the study of ancestry, is reviewed and criticized by Joseph Agassi in his *Towards an Historiography of Science,* pp. 31–45. Agassi calls it (p. 31) "the continuity theory," a theory whereby "all thinkers are greatly indebted to their predecessors, and all progress is made in small steps." He warns against the cancerous growth of continuity and says that historians of science should avoid relying on chronological relations without seeking causal connection.

Caverni listed, among other predecessors of Galileo, the sixteenth-century mathematician Giovanni Battista Benedetti (1530–1590) as one of Galileo's many precursors. Several authors, after Caverni, noted the similarity between the work of Benedetti and the early works of Galileo and suggested that Galileo had repeated and expanded Benedetti's work (see Giacomelli, *Galileo Galilei giovane e il suo "De motu"*). Koyré, in particular, believes that Galileo followed Benedetti step by step (*Galileo Studies,* pp. 21–61). Indeed, both Benedetti and Galileo emphasized the superior importance of theoretical foundations over experience in science. Both criticized Aristotle's theory of falling bodies in favor of the Archimedean theory of motion in media, and Galileo, in his *De motu,* even used the same terms, e.g., *impetus,* as Benedetti. Yet the apparent similarity between Galileo and Benedetti is explicable in many ways, and there is no evidence that Galileo relied on Benedetti in any way.

49. Wohlwill was soon accused by Favaro of being anti-Italian and anti-Latin, indicating that the whole discussion was also strongly influenced by the nationalistic feelings of that period (in less than a decade Italy would enter World War I against Austria and Germany). See Antonio Favaro, "Ancora, e per l'ultima volta, intorno all'episodio di Gustavo Adolfo di Svezia nei racconti della vita di Galileo," *Atti e memorie della R. Accademia di scienze lettere ed arti in Padova* (1906–1907) 23:6–12, p. 8.

50. Koyré, *Galileo Studies,* p. 65.

51. In his article, "An Experiment in Measurement," Koyré ridicules Galileo's description of the inclined plane experiment as being "an accumulation of sources of error and inexactitude" (p. 224).

52. See, in particular, Koyré's *Galileo Studies,* p. 37. The historical context of Galileo's Platonism, mainly the work of Iacopo Mazzoni, Galileo's colleague and friend at the University of Pisa, is studied by Paolo Galluzzi in "Il "platonismo" del tardo cinquecento e la filosofia di Galileo," in Paola Zambelli (ed.), *Ricerche sulla cultura dell'Italia moderna* (Bari: Laterza, 1973), pp. 39–79.

My outline here relies heavily on William Shea's presentation of Platonism and its relation to Galileo's experiments in *Galileo's Intellectual Revolution,* pp. 150–163. The function of thought experiment in Galileo's work seems to have been far more complicated than indicated by Koyré; see Gad Prudovsky, "The Confirmation of the Superposition Principle: On the Role of a Constructive Thought Experiment in Galileo's *Discorsi,*" *Studies in History and Philosophy of Science* (1980) 20:453–468.

53. Winifred L. Wisan, "Mathematics and Experiment in Galileo's Science of Motion," *AIMSSF* (1977) anno 2, fascicolo 2:149–160, p. 151.

54. The discussion is presented in Segre, "The Role of Experiment in Galileo's Physics."

55. Koyré's conclusion is strengthened, at least implicitly, by many later studies of Galileo's work; for instance, Ernest A. Moody, in "Galileo and Avempace: The Dynamics of the Leaning Tower Experiment," *Journal of the History of Ideas* (1951) 12:163–193, 375–422, offers more concrete evidence of Galileo's probable theoretical background and suggests the likelihood that Avempace was an origin of Galileo's work. More recently, the leading role of mathematics in Galileo's physics has been underlined by Winifred L. Wisan in "The New Science of Motion." Mathematics of course does not exclude experiment but, from Wisan's study, appears clearly to have been the more fundamental factor.

56. I owe this remark to Maurice A. Finocchiaro.

57. In chapter 5 of his *Galileo and the Art of Reasoning* (pp. 103–141), Finocchiaro has collected all the methodological remarks in Galileo's *Dialogues.*

58. *OG* 6:232; quotation from Drake, *Discoveries and Opinions,* pp. 237–238.

59. The fact that Galileo may not have been consistent in what he said has been pointed out by Paul K. Feyerabend in *Against Method.* Galileo's changing approach over time has been emphasized by Noretta Koertge in "Galileo and the Problem of Accidents," *Journal of the History of Ideas* (1977) 38:389–408.

60. See Maurice A. Finocchiaro, *Galileo and the Art of Reasoning,* chapters 5 and 7.

Chapter 3. Galileo's "Followers"

1. *OG* 19:628.

2. On the spreading of Newton's ideas see Ivo Schneider, *Isaac Newton* (Munich: Beck, 1988), chapter 7.

3. Joseph Agassi, in "Scientific Schools and their Success," *Science and Society,* pp. 164–191, claims that the existence of scientific schools is part and parcel of scientific activity, without, however, defining the concept.

4. Several modern collections of biographies deal with persons related to Galileo, in particular, Favaro, *ACG;* Abetti, *Amici e nemici.*

5. In 1978, a congress on the Galilean School was held in S. Margherita Ligure. One of the main lectures, by Luigi Bellone, concerned Marcello Malpighi (see G. Arrighi et al., *La scuola galileiana,* pp. 137–153). The contro-

versy over Milton's visit to Galileo is discussed by Neil Harris in "Galileo as Symbol: The 'Tuscan Artist' in Paradise Lost," *AIMSSF* (1985) anno 10, fascicolo 2 : 3–29, pp. 3–10.

6. E.g., L. Geymonat, in his *Galileo Galilei,* describes Galileo's cultural and political movement. M. L. Altieri Biagi, in *Galileo e la terminologia tecnico-scientifica,* a study of the language of Galileo and his followers, emphasizes their "esprit de corps" (p. 19).

7. Cavalieri, for instance, in 1621 (*OG* 13 : 55, 62) let Galileo know that he was trying to win some gentlemen in Milan to the Galilean cause, including Cardinal Borromeo (the founder of the Biblioteca Ambrosiana). Niccolò Aggiunti reported in 1627 (*OG* 13 : 357) on how he was "converting" Aristotelean students in Pisa. And Castelli (*OG* 13 : 338, 14 : 297, 359) kept Galileo well informed of his "missionary" work in Rome and other places.

8. In the nineteenth century, Castelli was suspected of having plagiarized the work from Leonardo da Vinci, but there is no evidence that Castelli knew Leonardo's theoretical hydraulics, and it is more likely that he reached similar results independently. Regrettably, many of his manuscripts were lost, some of them destroyed by the World War II bombing of the Benedictine abbey of Monte Cassino in the south of Italy. See Gino Arrighi, "Benedetto Castelli: considerazioni e proposte," *La scuola Galileiana,* pp. 3–11, especially p. 4, on the manuscripts lost in Monte Cassino, pp. 7–8 on Leonardo.

9. *OG* 18 : 303. Redondi describes the intellectual life of Rome during these years in *Galileo Heretic,* as does Maurizio Torrini in "Due galileiani a Roma: Raffaello Magiotti e Antonio Nardi" in G. Arrighi et al., *La scuola galileiana,* pp. 53–88. Castelli's book is discussed by Massimo Bucciantini in "Il trattato 'Della misura dell'acque correnti' di Benedetto Castelli. Una discussione 'sulle acque' all'interno della scuola galileiana." *AIMSSF* (1983) anno 8, fascicolo 2 : 130–140.

10. *OG* 14 : 387–388.

11. Caverni, *Storia del metodo sperimentale,* 1 : 176.

12. The life and work of Magiotti and Nardi is outlined by Maurizio Torrini in "Due galileiani."

13. *OG* 20 : 492.

14. According to Brissoni, "Antonio Nardi e Democrito," pp. 53–56, Nardi may have decided not to publish his writings because of moral scruples; his Democritan opinions contradicted his religious beliefs.

15. *Carteggio* 1 : 230.

16. Gal. MS 130, 739: "O quanto confuse sono queste academiche scene Parrebero l'idea della confusione se idea la confusione avesse. Ma se ordinate fossino non sarebbero formate da un confuso. Io però ne stimo che siano un caos filosofico, il quale facilmente ordinarsi si possa, purché la mente gli sopravvivi." ("How confused are these academic scenes! They would seem to give an idea of confusion, if confusion had an idea, but if they were to be ordered they would not be formulated by a confused person. I nevertheless believe that they are a philososophical chaos which can be easily ordered, provided that mind can survive it").

17. Gal. MS 130, 739–759. The detailed content of the *Scene* is presented by Procissi, *La collezione galileiana* 2:89–104.

18. Torrini, "Due galileiani," p. 85.

19. With Magiotti's help, Berti (preceding Torricelli) tried unsuccessfully to produce a vacuum by means of siphons. There is also evidence that he was interested in Copernicanism, even after Galileo's trial. His rank and his relations in Rome suggest that he was a priest—probably even a monk—although I found no documents proving it.

20. Caverni, *Storia del metodo Sperimentale,* 1:159.

21. I rely on Giusti's *Bonaventura Cavalieri and the Theory of Indivisibles,* which includes a detailed bibliography of Cavalieri's mathematical work (pp. 91–95). A more technical study of the theory of indivisibles is Kirsti Andersen's "Cavalieri's Method of Indivisibles."

22. Urbano D'Aviso, *Trattato della sfera* (Rome, 1682). I was able to see only a later edition of D'Aviso's life of Cavalieri, in his *Sfera astronomica* (Rome, 1690), pp. *xii–xxiv.*

23. *OG* 14:36.

24. Favaro wrote a biography of Aggiunti (*ACG* 3:1165–1243. See p. 1175). I found, however, no documents supporting Favaro's claim that the grand duke's decision was related to Galileo's recommendation.

25. Relevant correspondence is collected at the beginning of *OG* 14. Marsili's request is on p. 33; his original letter is in Gal. MS 19, 151.

26. Koestler, *The Sleepwalkers;* see, for instance, p. 378.

27. *OG* 14:386–387, 394–395, 411.

28. *OG* 18:189, 197.

29. *OG* 15, 277.

30. Gal. MS 11, 82.

31. Gal. MS 11, 64.

32. This is perhaps an outcome of the traditional inductivist approach in the history of science—strongly criticised by Joseph Agassi (*Towards an Historiography of Science*)—whereby ideas and even thinkers are either white or black. This approach is encouraged to some extent by the fact that Newton was born less than a year after Galileo's death, so that everything that came between Galileo's death and Newton's discoveries could easily be labeled as "black," or nonexistent.

33. *Carteggio* 1:6–8.

34. *OT* 4:23.

35. Caverni, *Storia del metodo sperimentale,* 1:181.

36. *Carteggio* 1:85; *OT* 4:81. The salary is given by Lanfranco Belloni in *Opere scelte di Evangelista Torricelli,* p. 17 (regrettably without quoting the source).

37. *OG* 19:626. In Torricelli's will, *OT* 4:91. Magnani's note is in Gal. MS 131, 46, published by Procissi in *La collezione galileiana* 2:106. The information about Torricelli's coffin is reported by Giovanni Ghinassi in *Lettere fin quì inedite di Evangelista Torricelli precedute dalla vita di lui* (Faenza, 1864), p. *L* (50 Roman). Torricelli's bones were however dispersed; see *OT* 4:111–120 and

Angelo Lama, "Per la ricerca delle ossa di Evangelista Torricelli," *Torricelliana,* pp. 71–78.

38. *OT* 4:23.

39. *Carteggio* 1:17.

40. Documents relating to Torricelli'a life and career are collected in volume 131 of the Galilean Collection of manuscripts.

41. Gal MS 131, 42.

42. Paolo Galluzzi, "Vecchie e nuove prospettive torricelliane," Gino Arrighi et al., *La scuola galileiana,* pp. 13–51, p. 46.

43. On the vicissitudes of Torricelli's manuscripts, see the Introduction to *OT* 1.

44. The description of the experiment is in a famous letter to Michelangelo Ricci, *Carteggio* 1:122–123. The quotation is from a translation by W. E. Knowles Middleton in *The History of the Barometer,* p. 23.

45. Redondi, *Galileo Heretic.* A history of the various philosophical and theological views of vacuum before Galileo is also presented by Edward Grant in *Much Ado about Nothing: Theories of Space and Vacuum from the Middle Ages to the Scientific Revolution* (Cambridge: Cambridge University Press, 1981).

46. *OT* 4:84–85. On Torricelli's astronomical work, see Domenico Benini, "La cultura astronomica di Evangelista Torricelli," *Torricelliana,* pp. 54–55.

47. Gilles Personier de Roberval, *Aristarchi Samii De Mundi Systemate* (Paris, 1644).

48. *Carteggio* 1:174–175, 204–205, 215–216, 275, 310–311 (quotation from p. 311). This exchange of letters is mentioned in Maurizio Torrini, *Dopo Galileo,* p. 13.

49. On Ferdinand's encouragement of science see, in particular, *TT* 1:147–150.

Chapter 4. The Indivisibles

1. Galileo's atomism has received relatively little study. Among more recent articles are William R. Shea, "Galileo's Atomistic Hypothesis," *Ambix* (1970) 17:13–27; A. Mark Smith, "Galileo's Theory of Indivisibles: Revolution or Compromise?," *Journal of the History of Ideas* (1976) 37:571–588; and H. E. Le Grand, "Galileo's Matter Theory," in *New Perspective on Galileo,* Robert E. Butts and Joseph C. Pitt (eds.) (Dordrecht and Boston: Reidel, 1978), pp. 197–208.

2. A thorough investigation of mathematics in Galileo's *Two New Sciences* is Wisan's "The New Science of Motion." The mathematics involved in the Law of Free Fall is outlined by Enrico Giusti in "Aspetti matematici della cinematica galileiana," *Bollettino di Storia delle Scienze Matematiche* (1981) 2:3–42.

3. Difficulties connected with atomism in the seventeenth century are illustrated by Christoph Meinel in "Early Seventeenth-Century Atomism."

4. Albert Einstein, in his foreword to Drake's translation of the *Dialogue* (pp. *xvii–xix*), remarked that "it was not until the nineteenth century that logi-

cal (mathematical) systems whose structures were completely independent of any empirical content had been cleanly extracted."

5. *OG* 1:252. Drabkin, p. 15.

6. *OG* 4:527. Translation from Drake, *Galileo at Work,* pp. 216–217.

7. *Risposta alle opposizioni del Sig. Lodovico delle Colombe, e del Sig. Vincenzio di Grazia* . . . (Florence, 1615), *OG* 4:441–789. See p. 732.

8. *Discorso delle comete di Mario Guiducci* (Florence, 1619), *OG* 6:37–108.

9. Lothario Sarsi, *Libra astronomica ac philosophica* . . . (Perugia, 1619), *OG* 6:109–180.

10. For example, Galileo corresponded with Giovanni di Guevara, the author of *Aristotelis Mechanicas commentarii* (Rome, 1627), distributed in 1629. Guevara's treatment of the problem of the "Wheel of Aristotle" in this work was later praised by Galileo in his *Two New Sciences.*

11. *OG* 13:309, 312, 318, 323.

12. *Esercitationi filosofiche di D. Antonio Rocco* . . . (Venice, 1633). Rocco's work was published by Favaro in volume 7 of the National Edition, together with the *Dialogue, OG* 7:596–750.

13. *OG* 7:682–683.

14. Galileo also addressed the problem of the "Wheel of Aristotle" (*OG* 7:68–72, Crew and de Salvio, pp. 20–25), praising the work of Giovanni di Guevara. But, as William A. Wallace points out in "The Problem of Apodictic Proof in Early Seventeenth-Century Mechanics: Galileo, Guevara, and the Jesuits," *Science in Context* (Spring 1989) 1:67–87, Galileo's solution was quite different from Guevara's. Wallace remarks (p. 75) that "Galileo was not so much interested in solving the problem posed by Aristotle as he was in using 'Aristotle's wheel' to show that there can be an infinite number of vacuums in a finite extent of matter."

15. For a concise presentation of Galileo's treatment, see Charles B. Boyer, *The History of the Calculus and Its Conceptual Development (The Concepts of the Calculus)* (New York: Dover, 1959), particularly pp. 115–117.

16. *OG* 8:72, Crew and de Salvio, p. 25. Italics are mine.

17. Ibid.

18. *OG* 18:72, Crew and de Salvio, p. 26.

19. *OG* 8:83, Crew and de Salvio, p. 38.

20. Later he even contradicted himself. In a letter written in 1640 to the philosopher Fortunio Liceti he denied ever having explained the essence of light (*OG* 18:208). Galileo, by the way, also maintained this ambivalent approach when dealing with infinitely large quantities, e.g., when he faced the question whether the universe is infinitely large. As Koyré remarks, in *From the Closed World to the Infinite Universe* (Baltimore and London: The Johns Hopkins Press, 1957, 1970), p. 95, Galileo "seems not to have made up his mind, or even, though inclining towards infinity, to consider the question as being insoluble."

21. Gal. MS 130, 971 (transcribed by Caverni in *Storia del metodo sperimentale* 4:138): "Ora il Galilei, benché parli molto ambiguo degl'infiniti di atto e

di potenza di numero e di mole, come anche del continuo e del congiunto, e di altri somiglianti principii, contuttociò si lascia intendere essere i corpi composti d'infiniti indivisibli attuali, . . . ma questi principii, con altre conseguenze, hanno bisogno di ridursi a' buon senso."

22. In fact a method like that sought by Cavalieri did exist. It is described in a manuscript discovered by J. L. Heiberg in 1906, in the form of a letter from Archimedes to Eratosthenes. See Thomas L. Heath (ed.), *The Method of Archimedes, Recently Discovered by Heiberg: A Supplement to The Works of Archimedes 1897)* (Cambridge: Cambridge University Press, 1912). On Archimedes's rigor in proof, see Ivo Schneider, *Archimedes* (Darmstadt: Wissenschaftliche Buchgesellschaft, 1979), pp. 43–63.

23. Cavalieri, *Geometria,* edited by L. Lombardo-Radice, p. 188.

24. Cavalieri proved this theorem for $n = 1$ and $n = 2$ in his *Geometria,* book II, theorems 19 and 24, respectively. He proved the remaining cases in his *Exercitationes.* Cavalieri's proofs, more or less in his own words and also in today's mathematical terms, are presented by D. J. Struik (ed.), *A Source Book in Mathematics, 1200–1800* (Cambridge, Mass.: Harvard University Press, 1969), pp. 214–219.

25. Gal. MS 130, 1010: "i principii ingegnosissimi degli indivisibli che a' giorni nostri incominciano a sorgere con una meravigliosa felicità."

26. *Carteggio* 1 : 1. Curved indivisibles are, according to Torricelli (*OT* 1-1 : 174), "In plane figures, the boundaries of circles, and in solid figures, the spherical, cyclindrical or conic surfaces which can adapt themselves perfectly to the figures and have (so to speak) a thickness which is always equal and uniform." One example is the comparison of a circle and a triangle (*OT* 1-1 : 174–175).

The ratio between the circumferences of the outer and inner (dashed) circles in figure N4.1 is also the ratio between their radii, which equals IL/BC. If the line AC is drawn so that BC equals the circumference of the outer circle, then the circumferences of all the "inner circles" OA equal the corresponding segments IL of their tangents parallel to BC within the triangle ABC, and each smaller circle corresponds to a smaller triangle AIL. Since the area of any circle equals half the product of its circumference and its radius, and the legs AI and IL of triangle AIL equal, respectively, the radius and the circumference of the smaller circle OA, then the area of triangle AIL ($AI \times IL/2$) is the same as that of the corresponding circle OA. In modern terms, curved indivisibles are to "straight" indivisibles as polar coordinates are to Cartesian coordinates.

In figure N4.2, Torricelli showed that the volume of the hyperbolic solid EBD equals the volume of the circular cylinder ACGH (where AH is twice AC), which is, of course, finite.

27. *OT* 1 : 191–213.

28. *Carteggio* 1 : 286, cf. 1 : 28–31. On Torricelli's work on centers of gravity, see Amedeo Agostini's "I baricentri trovati da Torricelli," *Bollettino della Unione Matematica Italiana* (1951) serie 3, anno 6, numero 2 : 149–159. Of

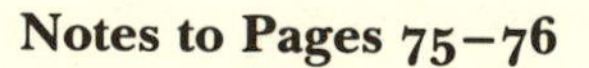

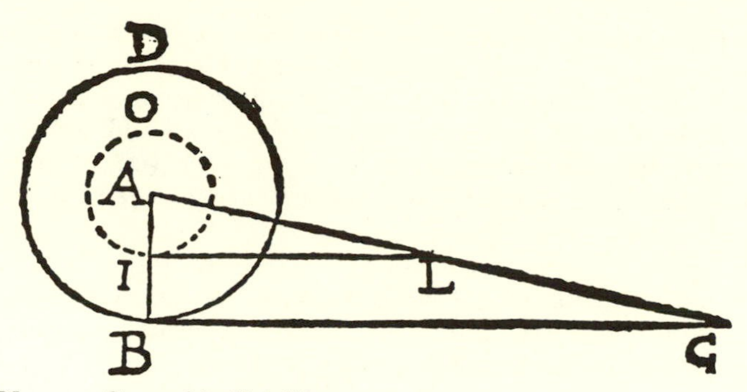

Figure N4.1. Curved indivisible.

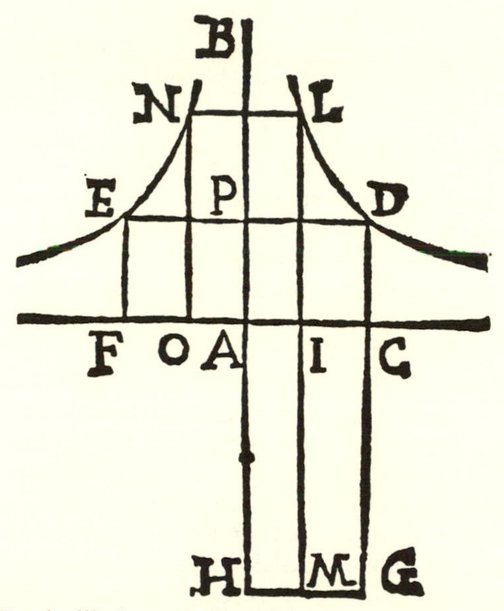

Figure N4.2. Torricelli's hyperbolic solid.

course, methods to find centers of gravity existed in ancient times. See Schneider, *Archimedes*, pp. 64–70, 109–124.

29. Guldin was born a Jew—his original name was Habakkuk Guldin—but was brought up as a Protestant and at the age of twenty converted to Catholicism.

30. He later taught at the Jesuit College in Graz and at the University of Vienna.

31. Bonaventura Cavalieri, *Trigonometria plana et sphaerica* (Bologna, 1643), *Carteggio* 1 : 37.

32. *Carteggio* 1:124. Torricelli gave demonstrations both new, using the theory of indivisibles, and classical, without using indivisibles. He showed, for instance, no fewer than twenty-one ways of squaring the parabola, ten by classical methods (*OT* 1, part 1:89–162). He compared himself to an artist presenting many different samples of his work (*OT* 1, part 1:95), also demonstrating by examples, that old and new geometry are perfectly equivalent.

33. *OT* 1, part 2:45–48, 415–426.

34. *OT* 1, part 2:320: "Che gli indivisibli tutti sieno eguali fra di loro, cioè i punti alli punti, le linee in larghezza alle linee e le superficie in profondità alle superficie, è opinione a giudizio mio non solo difficile da provarsi, ma anco falsa."

35. Gal. MS 130, 23: "L'universale ed intimo soggetto delle matematiche (cioè il quanto e i suoi modi e accidenti) cade sotto la metafisicale contemplazione, ne si può dall'humano ingegno quidditative almeno perfettamente conoscere, e quindi ci resta ignota la definizione propria e l'ultima ragione del continuo. Nasce questo dall'haver noi posto che i sensi non arrivano, se non all'esterno delle cose, e che non riduce nell'intelletto nostro intuiti[o]ne."

36. Gal. MS 130, 970: "In grazia de' matematici, ho posto accademicamente che le linee, i punti, e le superfici siano in atto ne' i corpi, come parti veraci e componenti da che nę seguirebbe l'haver quei termini propria esistenza e nulla vieterebbe potersi da qualche forza separarsi dal soggetto, e da qualche suprema potersi separar tutti. Quindi si darebbe uno spazio, et un numero infinito, il che ripugna alle cose poste." Transcribed by Caverni in *Storia del metodo sperimentale* 4:138.

37. In 1658, Pascal published a short treatise entitled *Histoire de la roulette* ("roulette" was his name for the cyloid; Roberval used the term trochoide), publicly accusing Torricelli of having plagiarized the theorem of the cycloid from Roberval. Pascal wrote that Roberval had demonstrated the theorem as early as 1634 and informed Mersenne who, in turn, communicated it to Galileo. Torricelli, according to Pascal, got hold of the theorem and published it ten years later as if it were his own discovery. Torricelli was defended by his friend Carlo Dati who, in his short treatise *Lettera a Filaleti,* produced documents to prove that Torricelli had reached his results alone. Dati quoted John Wallis, who had said that although Roberval, Descartes, and Fermat may have demonstrated the theorem, none of them published it, implying Torricelli's priority. Although these French mathematicians may have applied different standards than Torricelli and Wallis, the question went beyond standards of personal priority, and Dati was also protecting the priority of Italian science in general. In fact, French scientists had also often repeated, and some claimed priority for, Torricelli's barometer experiment. See *Oeuvres de Blaise Pascal,* vol. 8, edited by Leon Brunschvicg, Pierre Boutroux, and Felix Gazier (Paris: Hachette et Cie., 1886–1921), pp. 179–223. The history of the scandal was related by Ferdinando Jacoli in "Evangelista Torricelli ed il metodo delle tangenti detto Metodo del Roberval," *BB* (1875) 8:265–304.

Chapter 5. Torricelli's Rationale

1. For a full treatment, see chapter 2.

2. Cavalieri, *Lo specchio ustorio* (pages of the introduction are not numbered): "L'imparar Mercurio dalla testuggine di formar la lira, o dal batter vicendvol de' martelli l'inventar Pitagora la musica, fù un veder prima, in un certo modo, la conclusione, e da quella preconosciuta investigar poscia i suoi principii."

3. Ibid., "Ma il discorrer' il Colombo, che bisognava per tali, e tali ragioni, che vi fossero l'Indie nuove, e poi trovarle, fù un caminare da' principii alla conclusione, come per appunto par, che accada in questo proposito."

4. Ibid., "Dicendomi insieme, che chi era consapevole delle mie molte occupationi, m'havria scusato accettando volontieri per hora la parte della specolativa, per vederne poi con più commodità la cosa ridotta in prattica."

5. Koyré studied Galileo and Descartes and, to a lesser extent, Cavalieri and Torricelli in his *Galileo Studies;* Kepler and Borelli in *The Astronomical Revolution* and in "La mécanique céleste de J. A. Borelli," *Revue d'histoire des sciences* (1952) 5: 101–138; and Newton in his last work (published posthumously), *Newtonian Studies* (Cambridge, Mass.: Harvard University Press, 1965).

6. The broadest treatment of the barometer experiment is in de Waard, *L'expérience barométrique.* Two other relevant studies are Mario Gliozzi, "Origini e sviluppi dell'esperienza torriceliana," *OT* 4: 231–294, and W.E.K. Middleton, *The History of the Barometer,* chapter 2, pp. 19–54.

7. *OG* 1: 276–284, Drabkin, pp. 41–50.

8. *OG* 14: 157–160, p. 159: ". . . nell'aria, che siamo nel fondo della sua immensità, né sentiamo né il suo peso che la compressione che ci fa da ogni parte."

9. *OG* 8: 61–65, Crew and de Salvio, pp. 14–17.

10. *OG* 8: 61–62, Crew and de Salvio, pp. 14–15. Italics are mine.

11. Berti's experiment is described by de Waard in *L'expérience Barométrique,* pp. 101–110, and in English by Middleton in *The History of the Barometer,* pp. 10–18, and by Frank D. Prager in "Berti's Devices and Torricelli's Barometer from 1641 to 1643," *AIMSSF* (1980) anno 5, fascicolo 2: 35–53.

12. *OT* 4: 248.

13. Gliozzi admits this in his excellent article on Torricelli in *DSB* 13: 433–438, p. 438.

14. From de Waard, *L'expérience Barométrique,* pp. 178–181.

15. I rely on and quote Middleton's translations in *The History of the Barometer,* pp. 23–24.

16. Athanasius Kircher, *Musurgia universalis . . . ,* 2 vols. (Rome, 1650), vol. 1, pp. 11–13.

17. From de Waard, *L'expérience Barométrique,* pp. 182–184. On the equivocal role of experiment in science see Shapin and Schaffer, *Leviathan and the Air Pump,* which focuses on the related controversy between Boyle and Hobbes. On Boyle's interpretation of Torricelli's experiment, see pp. 40–49.

18. *OT* 1–2:472. There is no other evidence that such an experiment was carried out by Torricelli or Viviani, although the *Saggi*, published in 1667 (see chap. 8), indicate that the Accademia del Cimento performed many similar experiments.

19. Carl G. Hempel, *Aspects of Scientific Explanation and Other Essays in the Philosophy of Science* (New York, London: Free Press, Collier MacMillan, 1965), pp. 365–366, 454, and *Philosophy of Natural Science* (Englewood Cliffs, N.J.: Prentice-Hall, c. 1966), p. 9.

20. *OT* 1, part 2, p. 470: "Né il Torricelli incontrò a caso l'esperienza, ma guidato da un retto discordo, e nel tempo che vedde, e sperimentò l'effetto, avea già speculato la cagione."

21. *Carteggio* 1:125–127.

22. *Carteggio* 1:130–132.

23. I have used two later editions of these works: *Nova scientia* (Venice, 1562); *Quesiti et inventioni diverse* (Venice, 1554). The last work was reprinted with an introduction by Arnaldo Masotti (Brescia: Ateneo di Brescia, 1959). Selections of these works have been translated and annotated by Stillman Drake in *Mechanics in Sixteenth-Century Italy*, pp. 61–143.

24. Drake and Drabkin, *Mechanics in Sixteenth-Century Italy*, p. 63; Hall, *Ballistics*, p. 36.

25. *OG* 2:9.

26. Stillman Drake, "Tartaglia's Squadra and Galileo's Compasso," *AIMSSF* anno 2, fascicolo 1 (1977): 35–54.

27. *OG* 2:335–424.

28. See chapter 3.

29. Cavalieri, *Lo specchio ustorio*, pp. 151–172.

30. Ibid., p. 169.

31. However, Cavalieri may not have thought deeply enough. Unless the inertial component is somehow dissipated, the projectile will retain forward motion and miss the earth's center, going into a kind of orbit. (I owe this remark to Norman Rudnick.) Cavalieri did contemplate (on the same page) air resistance, which would diminish inertial motion, but I cannot tell if he intended this to apply to the present argument.

32. *OG* 8:272–274, Crew and de Salvio, pp. 248–250.

33. Galileo's distinction between slow and fast projectiles is discussed by Paul Lawrence Rose, "Galileo's Theory of Ballistics," *BJHS* (1968) 4: 156–159.

34. *OG* 18:125–126.

35. Transcribed in Caverni, *Storia del metodo sperimentale* 4:564–565. My transcription is directly from Gal. MS 130, 973–974 and varies slightly from Caverni's: "Due moti, quali insieme rimescolati non si mantengono ciascuni di essi sinceri, ma scambievolmente si alterano; e par necessario, che il violento sia più veloce nell'uscire dal proiciente, che allontanatone come anco nel natural moto avviene, e però non passerà di sua natural spazi eguali in tempi eguali (come in tal punto dubito il Galilei) anzi che alcuni mecanici si per-

suasero, che all'uscir la palla dall'artiglieria andasse per qualche spazio rettamente il che se bene vero non è (poiché non s'annulla l'azione della gravità) con tutto ciò par vero, che il moto orizontale sostenga da principio il proietto, sicché non discenda con la ragione, con quale discenderebbe per la sola gravità, ma nel progresso è necessario che la gravità vinca l'impeto straniero, acciò che si riconduca il proietto al centro, e così anche appar necessario che la forza, quale prima mosse un proietto dallo stato di quiete, non così muover lo possa, doppo l'acquisto e accrescimento di moto verso il centro. Ora dall'intrecciamento di questi moti, momenti, e tempi componsi una linea molto vicina alla parabolica, ma difficilissimo da me si reputa dimostrarla tale."

36. A short study of Torricelli's work on the motion of projectiles is by Amedeo Agostini, "Sul moto dei proietti di Evangelista Torricelli."

37. *OT* 2 : 101–249.

38. *OT* 2 : 209 f. in Italian.

39. *OT* 2 : 220.

40. The correspondence between Torricelli and Mersenne concerning Torricelli's "De Motu" is outlined and discussed in Galluzzi, "Evangelista Torricelli: Concezione della matematica."

41. *Carteggio* 1 : 208.

42. *OT* 3 : 349–356; Torricelli's answers, *OT* 3 : 381–392. Regrettably, this correspondence was not included in the *Carteggio*.

43. *Carteggio* 1 : 276: "Che i principii della dottrina *de motu* siano veri o falsi a me importa pochissimo. Poiché, se non son veri, fingasi che sian veri conforme habbiamo supposto e poi prendansi tutte le altre specolazioni derivate da essi principii, non come cose miste, ma pure geometriche. Io fingo o suppongo che qualche corpo o punto si muova all'ingiù et all'insù con la nota proporzione et horizontalmente con moto equabile. Quando questo sia, io dico che seguirà tutto quello che ha detto il Galileo et io ancora. Se poi le palle di piombo, di ferro, di pietra non osservano quella supposta proporzione, suo danno: noi diremo che non parliamo di esse."

44. *OT* 3 : 384: "Reiciamus ergo quicquid Physicum est in eo libello, nempe vocabula proiectorum, gravium, ballistarum etc. et Geometricum maneat, hoc est propositiones abstractae. Caeterum fabulae."

45. *Carteggio* 1 : 388: "Essendo pervenuta a Genova la sua opera del moto de' proietti, nella quale così al vivo si scuopre il suo acutissimo ingegno, diede occasione a questi nostri signori di farne diverse esperienze con il tiro di varie sorti di cannoni, e veramente mi ha reso assai stupefatto che tal teorica, così ben fondata, abbi così malamente risposto in pratica."

46. Segre, "Torricelli's Correspondence on Ballistics," pp. 493–494.

47. *Carteggio* 1 : 389: "Se l'autorità del Sig.r Galileo, del quale conviemmi esser parziale, non mi facesse resistenza, non mancherei d'avere qualche dubbio circa il moto de' proietti, se fusse parabolico o no, oppure, se è tale, non saprei accertarmi se l'asse della detta parabole debba essere perpendicolare all'orizonte o no."

48. *Carteggio* 1 : 391: "Non fù mai mia intenzione, quando io scrissi quel li-

bretto del moto, di voler sostener le cose che io affermavo in esso, se non *ex hypothesi,* cioè contro coloro i quali mi concederanno per vere quelle due famosissime supposizioni: che le discese del grave in tempi eguali siano *ut numeri impares ab unitate;* e che gli spazzi passati orizontalmente in tempi eguali siano eguali tra di loro."

49. *Carteggio* 1:391–392: "Molte di queste cause, le quali possono far discordare l'esperienza dalla dimostrazione, furono avvertite dal Galileo nel suo libro del moto. La principalissima però è l'impedimento dell'aria, la quale resiste ad ogni sorte di moto, ma molto più quando il mobile sarà più veloce; et in somma moltissimo quando il proietto verrà cacciato dalla furia soprannaturale del fuoco, che è la massima di tutte le nostre velocità e naturali et artificiali. Non è però maraviglia se l'esperienze, particolarmente quelle che si fanno con macchine da fuoco, riescano diverse dalla dimostrazione. Il supposto nostro è che l'impeto orizontale si mantenga sempre il medesimo; la pratica però dimostra che l'impeto orizontale presso la bocca delle macchine è quattro e sei volte maggiore che presso alla fine del tiro."

50. Hall, *Ballistics,* p. 16.

51. *Carteggio* 1:399–400.

52. *Carteggio* 1:405–407.

53. Hall, *Ballistics,* pp. 98–99.

54. In Torricelli, *Opere scelte,* Belloni (ed.) pp. 29–36; see p. 31.

55. Galluzzi, "Evangelista Torricelli. Concezione della matematica."

56. Torricelli's letter to Galileo, *OG* 14:387.

57. Regrettably, the editors of the *OT* artificially separated Torricelli's mathematical works (vol. 1) from his "physical" ones (vol. 2).

Chapter 6. Viviani's Hesitations

1. Viviani reported details of his life on many occasions, for instance, in a long autobiographical letter in 1697 to the Abbé Marquis Salviati, published in Fabroni, *Lettere* 1:4–22. The broadest outline I found of Viviani's life is Favaro's "Vincenzio Viviani," in *ACG* 2:1007–1163. A more recent biography—also one of the few that do not unduly venerate—is Maria Luisa Bonelli's, "L'ultimo discepolo: Vincenzio Viviani," in Carlo Maccagni (ed.) *Saggi su Galileo Galilei* (Florence: Barbèra, 1972), pp. 656–688. This article relies on some of Viviani's unpublished manuscripts but still does not cover all his work. An outline in English of Viviani's life and work in relation to his contribution to the Accademia del Cimento is presented in Middleton, *The Experimenters,* pp. 36–39.

2. *Carteggio* 2:96.

3. Fabroni, *Lettere* 1:11.

4. The documents of Viviani's official discharge were published by Fabroni in *Lettere* 1:22–24.

5. See *ACG* 2:1046–1055.

6. Gal. MS 155, 1–4.

7. *ACG* 2:1095.

8. Gal. MSS 155 to 258. The contents of Viviani's manuscripts are outlined in *ACG* 2:1134–1155.

9. The relations between Viviani and Borelli are described in Middleton, *The Experimenters*, pp. 310–316.

10. *De maximis et minimis geometrica divinatio in quintum Conicorum Apollonii Pergaei adhuc desideratum* . . . (Florence, 1659).

11. Borelli's translation appeared under the title *Apollonii Pergaei conicorum libri V. VI. et VII* . . . (Florence, 1661).

12. A complete bibliography of Viviani is presented by Pietro Riccardi, *Biblioteca matematica italiana* . . . , 2 vols. (Milan: Goerlich, 1952), 1:625–630. One of Viviani's achievements was proposing and solving the "Florentine enigma" (see A. Natucci, "Vincenzo Viviani," *DSB* 14:49).

13. *DSB* 14:49.

14. This incident was described by Settimi in a letter to Grand Duke Ferdinand II, *OG* 18:372.

15. Favaro, *ACG* 2:1119.

16. A list of Viviani's correspondents was compiled by Favaro in *ACG* 2:1156–1163.

17. Jarvie's term is reported in Joseph Agassi, *Towards a Rational Philosophical Anthropology* (The Hague: M. Nijhoff, 1977), p. 3.

18. Bonelli, "L'ultimo discepolo," p. 687.

19. Middleton, *The Experimenters*, p. 37.

20. Ibid., p. 29.

21. A. Favaro, "Documenti inediti per la storia dei manoscritti galileiani nella Biblioteca Nazionale di Firenze," *BB* (1885) 18:1–112, 151–230, pp. 17–18.

22. Viviani's numerous enquiries concerning Galileo's life are evident in the many documents and letters collected in Gal. MS 11. The detailed contents of this volume are presented by Procissi in *La collezione galileiana*, 1:13–16. Information about the bust and the inscription was given by Viviani in his *Divinatio in Aristei* . . . (Florence, 1701), pp. 121–128 and related diagrams.

23. *Carteggio* 2:38–39.

24. Relevant correspondence is found in *Carteggio* 2.

25. *Carteggio* 2:302.

26. The Bologna edition appeared under the title *Opere di Galileo Galilei. . . . In questa nuova edizione insieme raccolte, e di varrii Trattati dell'istesso autore non più stampati accresciute* (Bologna, 1655–1656). Viviani's complaint is in *Carteggio* 2:321.

27. Amply attested by Favaro's "Documenti inediti."

28. The earliest published biography of Galileo may have been written by Thomas Salusbury in the second part of the second volume of his *Mathematical Collections*, published in 1665. Unfortunately, no copies of this part have survived. See Stillman Drake, "Galileo in English Literature of the Seven-

teenth Century," in McMullin (ed.), *Galileo: Man of Science,* pp. 415–431, especially pp. 426–427. I am indebted to Professor McMullin for having drawn my attention to this biographical "ghost."

Chapter 7. Patterns of a Renaissance Biography

1. *OG* 19:606.
2. *OG* 19:629.
3. See chapter 2.
4. *OG* 19:638.
5. *OG* 19:642.
6. The handwritten copy of Gherardini's life of Galileo is in Gal. MS 11, 3–19. Viviani's annotations follow in folio 20. Vincenzio Galilei's notes are in Gal. MS 11, 126–129 and have been published in *OG* 19:594–596.
7. *OG* 19:647–659. A draft of this "letter" is in Gal. MS 85, 39–50. (A second draft of the essay is in the National Library of Paris.)
8. *A:* Gal. MS 11, 72–118. *B:* Gal. MS 11, 22–68.
9. The history of Viviani's life of Galileo is told by Antonio Favaro in "Vincenzio Viviani e la sua 'Vita di Galileo'," *Atti del R. Istituto Veneto di Scienze, Lettere ed Arti,* vol. 62, part 2 (1902–1903): 683–703.
10. Maximilien Marie, for instance, in *Histoire des sciences mathématique et physique,* 12 vols. (Paris, 1883–1888; reprinted, New York, 1977), 4:90, proposes giving Cavalieri the prize for obscurity. Alexandre Koyré, in "La mécanique de J. A. Borelli," p. 101, says that Borelli wrote atrociously, discouraging the best-willed reader. Galileo's writing style, with particular reference to his newly created scientific terminology, is studied in Maria Luisa Altieri Biagi, *Galileo e la terminologia tecnico-scientifica.*
11. Furio Diaz, *Il granducato di Toscana,* pp. 422–463, particularly pp. 424–427.
12. *Carteggio* 1:18 (italics are mine): "Sento che vogliono cose piuttosto fisiche che mattematiche, e forsi con ragione, poiché quelle assomiglierei io piuttosto alla crusca, e queste al fior di farina, vero cibo e nutrimento dell'intelletto. Nondimeno conviene accomodarsi al loro genio, anzi al genio universale, che non istima punto le mattematiche se non ne vede qualche applicazione."
13. Ibid.: "Onde conviene esser fornito di due sorti di roba per soddisfare a tutti i gusti. Anzi per soddisfare al pubblico, che argomenta il valore de' dottori e delle dottrine dal numero de' seguaci, bisogna provvedersi di quella che è più di spaccio e per servirlo bene ingannarlo, o direi assassinare gl'ingegni, poiché il pubblico vuole esser trattato così per esser servito bene."
14. Strictly speaking, as Maurice A. Finocchiaro pointed out to me, one should distinguish between empiricism and practical applications, and Cavalieri's remark is perhaps more related to the latter. Yet Viviani and his contemporaries do not seem to have paid much attention to this distinction. For them it seems to have been more important for science to appear "tangible"

rather than abstract, and both the empirical and practical aspects of science satisfied this requirement; in fact, Viviani spoke of both aspects.

15. Baldi's manuscript is now kept at the Centro Internazionale di Studi Rosminiani in Stresa, Italy. Selections were published in Moritz Steinschneider, "Vite di matematici arabi, tratte da un'opera inedita di Bernardino Baldi," *BB* (1872) 5:427–534, and in Enrico Narducci, "Vite inedite di matematici italiani scritte da Bernardino Baldi," ibid. (1886) 19:335–406, 437–489, 521–640.

16. Rose, *The Italian Renaissance of Mathematics,* pp. 253–279. On Baldi's work see also P. L. Rose, "Copernicus and Urbino: Remarks on Bernardino Baldi's *Vita di Niccolò Copernico* (1588)," *Isis* (1974) 65:387–389.

17. The development of historiography during the Renaissance and in Viviani's day is treated in many works, for instance, Eric Cochrane, *Historians and Historiography in the Italian Renaissance* (Chicago, London: University of Chicago Press, 1981), especially chapter 14, dedicated to biography, pp. 393–422. A classical survey of the sources of the modern history of art is Julius von Schlosser, *Die Kunstliteratur* (Vienna: Schroll, 1924), which was available to me in the Italian version: Julius Schlosser Magnino, *La letteratura artistica. Manuale delle fonti della storia dell'arte moderna,* translated by Filippo Rossi, third edition revised by Otto Kurz (Florence and Vienna: La Nuova Italia and Schroll, 1964). A concise account of the evolution of Vasarian historiography during the seventeenth century is presented by Martino Capucci, "Dalla biografia alla storia. Note sulla formazione della storiografia artistica nel seicento," *Studi Secenteschi* (1968) 9:81–125. On the same subject, Capucci, "Forme di biografia nel Vasari," in *Il Vasari storiografo e artista. Atti del congresso internazionale nel IV centenario della morte* (Florence: Istituto Nazionale di Studi sul Rinascimento, 1976), pp. 299–320; Ferruccio Ulivi, "L'eredità del Vasari in Italia," ibid., pp. 525–532.

18. Carlo Cesare Malvasia, *Le pitture di Bologna* (Bologna, 1686. Reprinted, Bologna: Alfa, 1969), pp. 1–2. Malvasia also falsified evidence; see Marcella Brascaglia's Introduction to the reprint of Malvasia's *Felsina pittrice: vite dei pittori bolognesi* (Bologna, 1678. Reprinted, Bologna: Alfa, 1971), p. 15.

19. Capucci, "Dalla biografia alla storia," p. 108. Malvasia does not appear in the list of Viviani's correspondents published by Favaro in "Vincenzio Viviani," *ACG,* pp. 1156–1163.

20. P. 3; the italics are mine: "A me basterà il solo guidarvi ove possiate rendervene capace colla semplice occulare ispezione. L'evidenza di fatto esser deve sol quella, che ne costituisca oggi voi giudice; & a somiglianza all'odierne sperienze della non meno tanto rimota Inghilterra, che della prossima a noi Firenze." I am indebted to Gabriele Bickendorf for pointing out this significant passage to me.

21. Filippo Baldinucci, *Notizie dei professori del disegno da Cimabue in qua . . . ,* 6 vols. (Florence, 1681–1728). A later edition, of 1845–1847, was reprinted in 8 volumes (Florence: Studio per ed. scelte, 1974–1975). On Baldinucci and Prince Leopold see Paola Barocchi, "Il collezionismo del Cardinale Leopoldo

e la storiografia di Baldinucci," *Omaggio a Leopoldo de' Medici* (Florence: Olschki, 1976), pp. 14–25.

22. The quotation on folio 168*r* is taken from Vasari's *Vite* (p. 774 of the 1568 edition). Vasari, *La Vita di Michelangelo,* Barocchi (ed.), 1 : 116. It reads as follows:

> Il Vasari nel libro della Vita di Michelangelo stampato da Giunti in Firenze nel 1568 in 4° a faccia *774* così, dopo lunghi racconti, scrisse di lui "con conoscimento grandissimo fece testamento di tre parole che lasciava l'anima sua nelle mani di Dio, il suo corpo alla terra; e la roba a' parenti più prossimi imponendo a' suoi che nel passar di questa vita gli ricordassino il patire di Gesù Cristo. E così a 17 di febbraio l'anno 1563, a ore 23 a uso Fiorentino, che al Romano sarebbe *1564* spirò per irsene a miglior vita."

This may be translated as follows:

> Vasari, in the book of Michelangelo's life, printed in 1568 in Florence by Giunti in 4°, folio 774, after long tales, wrote of him: "With perfect consciousness he made his will in three sentences, leaving his soul to God, his body to the earth, and his material possessions to his nearest relations, asking his friends that as he died they should recall to him the sufferings of Jesus Christ. And so on 17 February, in the year 1563, at the twenty-third hour according to Florentine reckoning, which by the Roman is 1564, he breathed his last and went to a better life."

I have relied heavily on Vasari, *Lives of the Artists,* translated by Bull, 1 : 417. Vasari reported, mistakenly, that Michelangelo was born on February 17; see Vasari, *La Vita di Michelangelo,* Barocchi (ed.), 4 : 1834–1835. As Gal. MSS 168v and 171, and also his *Divinatio Aristaei* (p. 127), testify, Viviani knew that Vasari was mistaken.

The quotation on 168v relies on the same paragraph in Vasari, which reads as follows:

> Vincenzio Viviani riversisce il Sig. Filippo Baldinucci e supplica a favorirlo quì sotto della notizia dell'anno, mese, giorno, ora, e luogo della morte del divino Michelangelo Buonarroti.
>
> Ricordo cavato dal Vasari
> Il divino Michelangelo Buonarroti nella città di Roma ammalatosi d'una lenta febbre alli 17 di febbraio dell'anno 1563 alle 23 ore, all' uso fiorentino
> passò da questa all'altra vita.
> all'uso romano l'anno 1564.
> Il natalizio di questo grand'uomo fù dell'anno 1474 il giorno 6 di marzo
> in domenica alle 8 di notte al fiorentino.

This may be translated as follows:

> Vincenzio Viviani pays his respects to Filippo Baldinucci and begs him to write down here the year, month, day, hour and place of the death of the divine Michelangelo Buonarroti.

Recollection taken from Vasari
The divine Michelangelo Buonarroti in the city of Rome fell ill with a slow fever
[and] on the 17th of February 1563 at the twenty-third hour
passed from this to the other life.
According to the Roman reckoning, the year 1564.
The birth of this great man was in the year 1474, on Monday 6th of March
at 8 at night in the Florentine style.

23. Kris and Kurz, *Legend, Myth, and Magic.* Kris's interest in psychoanalysis and contacts with Sigmund Freud certainly contributed to this research.

24. In Dante's *Purgatorio;* see Vasari, *Le Vite,* Bettarini (ed.), vol. 2 of the annotations, pp. 348–350.

25. *DSB* 5:237–250, especially p. 237. Viviani also reported the existence of links between Galileo and other leading painters of his age: "Cigoli" (Lodovico Cardi), "Bronzino" (Cristofano Allori), "Passignano" (Domenico Cresti da Passignano), and "Empoli" (Jacopo Chimenti): *OG* 19:602. I also found in Gal. MS 11 a document—folio 163v—stating the same thing, but I cannot say by whom and when it was written, or whether it is reliable.

26. Nelli, *Vita e commercio,* pp. 21–22.

27. *OG* 19:635.

28. Favaro, "Sul giorno della nascita di Galilei."

29. Ibid., p. 706.

30. I rely on A. Cappelli, *Cronologia, cronografia e calendario perpetuo* (Milan: Ulrico Hoepli, 1969), p. 11. A related document (Gal. MS 11, 147r) was published by Antonio Favaro in "Scampoli Galileiani 44: il matrimonio dei genitori di Galileo," *Atti e Memorie della R. Accademia di Scienze e Lettere ed Arti in Padova* anno 293, nuova serie, vol. 8 (1891–1892): 12–22, especially pp. 17–18.

31. The related documents are in Gal. MS 11, pp. 182–188. Cf. Favaro, "Sul Giorno," pp. 708–709. Galileo's certificate of baptism was published by Favaro in *OG* 19:25.

32. Virginia Galilei was born on August 12, 1600, and was baptised on August 21 (*OG* 19:218).

33. I owe this remark to Heribert M. Nobis.

34. *OG* 19:601: "Cominciò questi ne' prim'anni della sua fanciullezza a dar saggio della vivacità del suo ingegno, poiché nell'ore di spasso esercitavasi per lo più in fabbricarsi di propria mano varii strumenti e macchinette."

35. Vasari's statement is on p. 139 of the 1568 edition, Bettarini (ed.), 2:96.

36. *A:* Gal. MS 11, 77. *B:* Gal. MS 11, 28. Cf. *OG* 19:602.

37. *OG* 19:32.

38. *A:* Gal. MS 11, 80. *B:* Gal. MS 11, 31. Cf. *OG* 19:604.

39. *OG* 19:636–637.

40. *A:* Gal. MS 11, 81. *B:* Gal. MS 11, 32. Cf. *OG* 19:605.

41. *A:* Gal. MS 11, 82. *B:* Gal. MS 11, 33. Cf. *OG* 19:605. The Appendix was published in *OG* 1:179–208, among Galileo's early works. Galileo mentioned this treatise in a 1636 letter to Elia Diodati (*OG* 16:523–524). As told

in the *OG,* the original letter was lost, but the National Library in Florence owns several excerpts (Gal. MS 76, folios 85*r*, 76*v*, 29*r*, 73*r*, 147*r*). Diodati may have sent the original to Viviani, who copied it or had it copied. Two of these excerpts (85*r* and 76*r*) say that Galileo was 22 years old; the others say he was 21.

42. *OG* 19:602: "dalla natura fu eletto per disvelare al mondo parte di que' segreti."

43. See Antonio Favaro, "L'episodio di Gustavo Adolfo di Svezia nei racconti della vita di Galileo," *Atti del Reale Istituto Veneto di Scienze Lettere ed Arti,* vol. 65, part 2 (1905–1906): 17–39; "Ancora, e per l'ultima volta, intorno all'episodio di Gustavo Adolfo di Svezia nei racconti della vita di Galileo," *Atti e memorie della R. Accademia di Scienze Lettere ed Arti in Padova* anno 366, nuova serie, 23 (1906–1907): 6–12.

44. *OG* 19:606: "con salde dimostrazioni e discorsi."

45. For a broader treatment of the question see Segre, "Galileo, Viviani and the Tower of Pisa."

46. Galileo, *On Motion,* translated by Drabkin, pp. 27, 31 (note), 38, 87, 101, 107, 127; *Mechanics,* translated by Drabkin, p. 374.

47. *OG* 4:242.

48. *OG* 18:305–306.

49. Abetti and Pagnini, *Le Opere dei Discepoli di Galilei.*

50. *OG* 19:606. The italics are mine. *A:* Gal. MS 11, 83: "che tutto si vede poi diffusamente trattato da lui nelli ultimi Dialoghi delle due Nuove Scienze." *B:* Gal. MS 11, 34: "che tutto si vede poi diffusamente trattato da lui nelli suddetti Dialoghi delle Nuove Scienze."

51. *OG* 8:107–108. Crew and de Salvio, pp. 62–63. On this occasion Galileo showed that Aristotle's law leads to a logical contradiction. If a large stone falls faster than a smaller one, then in a union of the two the smaller one should slow the larger. But the two together form a larger stone than either alone, which, according to Aristotle, should fall faster. Hence Aristotle's law is self-contradictory.

52. Viviani however ended the story with a remark about the Aristotelian relationship between speed and the density of the medium, saying explicitly that Galileo refuted it by "inferring from the obvious absurdities against the senses which would follow as a consequence of it" ("inferendolo da manifestissimi assurdi ch'in conseguenza ne seguirebbero contro al senso medesimo"): *OG* 19:606.

Chapter 8. The Accademia del Cimento

1. See Fabroni, *Lettere* 1:13.

2. Ibid.: "Eh vi ho assegnato quella provvisione come a Lettore di Matematica, e non perché la legghiate. Non legge il Redi, non legge il Dati: queste son letture onorarie, che noi le diamo per aiuto a quelli, che son buoni a scrivere: quand'avrete qualche cosa all'ordine per la stampa ditemelo, ch'io farò conto d'aver un lettor di più a Pisa. Quei leggeranno a pochi presenti, e

voi scriverete a tutti, presenti e futuri. Scriverete cose, e cose vere, ed essi diranno parole che il vento poi se le porta."

3. *TT* 1:149–150. Cf. Middleton, *The Experimenters,* p. 21.

4. *TT* 1:151–156.

5. Gal. MSS 259–307.

6. The meaning of the term is explained by Middleton in *The Experimenters,* pp. 50–52.

7. Par. III, lines 1–3: "Quel sol che pria d'amor mi scaldò il petto // di bella verità m'avea scoverto, // provando e riprovando, il dolce aspetto" ("That sun which first had warmed my heart with love // Had now, by argument and refutation/Revealed to me the lovely face of truth"). Both quotations are taken from Middleton, ibid., p. 52. The English translation is by Lawrence Grant White (New York, 1948), p. 132.

There is a basic difference between two ways of translating the Accademia's motto. One translation, "testing and retesting," may indicate a totally experiential empiricism; a second, "testing and refuting" (or, to use Middleton's alternative expression, ibid. p. 53, "trial and error"), indicates a more experimental, a priori empiricism. All the English translations of the *Divine Comedy* I could find translate the word "riprovando" as "refuting." The term "refutation" has acquired a specific significance in the philosophy of science since Karl R. Popper published his two books, *Logik der Forschung: zur Erkenntnistheorie der modernen Naturwissenschaft* (Vienna: Springer, 1935), translated into English as *The Logic of Scientific Discovery* (London: Hutchinson, 1959), and *Conjectures and Refutations: The Growth of Scientific Knowledge* (London: Routledge and Kegan Paul, 1963, 1965). Popper maintained that scientific theories do not originate from experience but are created through an imaginative process, and that observation cannot verify a theory, but can at best serve to refute it. Are we therefore to conclude that the Accademia del Cimento is a confirmation of Popper's view and had an a priori approach to scientific investigation? Bruno Nardi's article "Significato del motto 'provando e riprovando,'" in *Celebrazione dell'Accademia del Cimento,* pp. 71–79, tends to interpret *provando e riprovando* as empirical and a posteriori. Observational empiricism, or modern a priori experimentalism? Such a modern reconstruction can hardly tell us what really went on in the Accademia.

8. I rely heavily on Middleton's translation, *The Experimenters,* p. 90.

9. *Carteggio* 2:312–313.

10. For a detailed biography of Oliva see Ugo Baldini, "Un libertino accademico del Cimento: Antonio Oliva," Supplemento agli *AIMSSF* anno 1977, fascicolo 1, monografia n. 1.

11. See Middleton, *The Experimenters,* p. 34.

12. *TT* 1:418.

13. *The Correspondence of Henry Oldenburg,* edited and translated by A. Rupert Hall and Marie Boas Hall, vol. 4 (Madison: University of Wisconsin Press, 1967), p. 248.

14. Gal. MS 259–270.

15. Gal. MS 296–307.

16. Abetti and Pagnini, *Le Opere dei discepoli,* p. 54. Gal. MSS 271–274. Part of these works were published in *TT* 2:737–800.

17. In addition to the documents published in *TT* 2-2:737–746 are studies of the works by A. Van Helden in "The Accademia del Cimento and Saturn's Ring, *Physis* (1973) 15:237–259.

18. *Eustachii Divini Brevis annotatio in systemata saturnium Christiani Eugenii* (Rome, 1660): *Oevres complètes de Christian Huygens,* vol. 15 (The Haage: Martinus Nijhof, 1925), pp. 403–437.

19. Fabroni, *Lettere* 1:120.

20. The controversy between Galileo and Grassi is also described and discussed thoroughly by Borelli in a long letter defending Galileo's views. See W. E. Knowles Middleton, "Some Unpublished Correspondence of Giovanni Alfonso Borelli," *AIMSSF* (1984) anno 9, fascicolo 2:99–132, pp. 109–115.

21. Gal. MS 272, 115–116.

22. Middleton, *The Experimenters,* p. 253.

23. Ibid., p. 4.

24. Gal. MS 275, 84–85. Dated November 19, 1657.

25. Middleton, *The Experimenters,* p. 33.

26. Gal. MS 275, 109–110: "Che da qualsivoglia grado di freddo non solo non si accresce la mole di alcuni fluidi, ma per il contrario sempre si diminuisce, questi sono l'acqua arzente, e l'argento vivo; e non ha' dubio, che se l'aumento dell'acqua addiacciata dependesse da intrusione di quei corpi loro frigorifici, dovrebbe necessariamente accrescendosi lo sforzo del freddo aumentarsi ancora la mole dell'argento vivo, e dell'acqua arzente, e non dovrebbe andar continuamente scemando; e per il contrario, supposto che il raffreddamento altro non sia che diminuzione di qui corpi ignei, necessariamente al maggior sforzo del freddo deve conseguire maggior diminuzione di mole, ma questo è conforme l'esperienza, e quello no; adunque il freddo è privazione di caldella."

27. Abetti and Pagnini, *Le Opere dei discepoli,* p. 252; Middleton, *The Experimenters,* pp. 246–247.

28. W. E. Knowles Middleton, "Carlo Rinaldini and the Discovery of Convection in Air," *Physis* (1968) 10:299–305.

29. Ibid., p. 303.

30. Ibid.

31. "Dissertazione di Carlo Dati sull'utilità, e diletto che reca la geometria," *TT* 2:314–327.

32. Described in detail by Shapin and Schaffer, *Leviathan and the Air-Pump.*

33. Torricelli, *Opere,* Belloni (ed.), p. 31.

34. To the best of my knowledge the only research carried out so far in this direction is by Adriano Prosperi, "L'Inquisizione fiorentina al tempo di Galileo," in Paolo Galluzzi (ed.), *Novità celesti e crisi del sapere: Atti del convegno internazionale di studi galileiani,* Supplemento agli *Annali dell'Istituto e Museo di Storia della Scienza* (Florence: Giunti-Barbèra, 1984). Prosperi consulted the archives of the Florentine Inquisition and found that many documents possi-

bly relevant to Galileo had disappeared. The remaining material on Galileo is too slight to justify any conclusions.

35. *TT* 1:399.

36. *TT* 1:124.

37. Favaro, "Documenti inediti," pp. 56–57.

38. Paolo Galluzzi, "L'Accademia del Cimento," p. 823.

39. Borelli, under the pseudonym of Pier Maria Motulo, *Lettera del movimento della cometa.* The work on Jupiter's satellites was *Theoricae mediceorum Planetarum.*

40. *TT* 1:385.

41. As elaborated by Finocchiaro, *Galileo and the Art of Reasoning,* especially chapter 7.

42. In a letter published by Tullio Derenzini in "Alcune lettere di Giovanni Alfonso Borelli ad Alessandro Marchetti," *Physis* (1959) 1:224–243, especially p. 233.

Epilogue

1. Fabri, *Brevis annotatio.*

2. See Redondi, *Galileo Heretic,* p. 319.

3. "Niuno de' Professori della sua Università di Pisa si legga e s'insegni pubblicamente né privatamente in scritto o in voce la filosofia Democratica, ovvero degli atomi, ma solo l'Aristotelica, e chi in modo alcuno cotravenisse alla volontà di S. A. oltre la rigorosa indignazione dell' A. S. s'intenda ipso facto licenziato." Quotation from Angelo Fabroni, *Historiae academiae Pisanae* (Pisa, 1795), vol. 3, pp. 410–411.

4. Ibid., pp. 319–320.

Bibliography

Abetti, Giorgio. *Amici e nemici di Galileo.* Milano: Bompiani, 1945.

Abetti, Giorgio, and Pagnini, Pietro, eds. *Le Opere dei Discepoli di Galileo Galilei: Edizione Nazionale.* Vol. I: *L'Accademia del Cimento.* Parte Prima. Florence, Barbèra: 1942.

Accademia del Cimento. *Saggi di naturali esperienze. . . .* Florence, 1667.

Agassi, Joseph. *Towards an Historiography of Science.* The Hague: Mouton, 1963; Middletown, Conn.: Wesleyan University Press, 1967.

———. *Science and Society: Studies in the Sociology of Science.* Dordrecht, Boston, London: Reidel, 1981.

Agostini, Amedeo. "Sul moto dei proiettili di Evangelista Torricelli." *Pubblicazioni Scientifiche a cura dell'Accademia Navale* N. 11. Leghorn, 1951.

Altieri Biagi, Maria Luisa. *Galileo e la terminologia tecnico-scientifica.* Florence: Olschki, 1965.

Altieri Biagi, Maria Luisa, ed. *Scienziati del seicento.* Milan: Rizzoli, 1969.

Andersen, Kirsti. "Cavalieri's Method of Indivisibles." *Archive for History of Exact Sciences* (1985) 31:291–367.

Arrighi, G., Galluzzi, P., Torrini, M., De Angeli, E., Baldini, U., and Belloni, L. *La scuola galileiana: Prospettive di ricerca.* Florence: La Nuova Italia, 1979.

Baldini, Ugo. "Giovanni Alfonso Borelli e la rivoluzione scientifica." *Physis* (1974) 26:97–128.

Biagioli, Mario. "Galileo the Emblem Maker." *Isis* (1990) 81:230–258.

Boffito, Giuseppe. *Bibliografia Galileiana (1896–1940).* Rome: Libreria dello Stato, 1943.

Borelli, Giovanni Alfonso. *Lettera del movimento della cometa apparsa il mese di dicembre del 1664.* Pisa, 1665.

———. *Theoricae mediceorum planetarum ex causis physicis deducta.* Florence, 1666.

Brissoni, Armando. "Antonio Nardi e Democrito: un pretesto per parlar di Galileo." *La Critica Politica* (January–March 1990) 14-1:37–70.

Burstyn, Harold L. "The Deflecting Force of the Earth's Rotation from Galileo to Newton." *Annals of Science* (1965) 21:47–80.

Carugo, Adriano, and Crombie, Alistair C. "The Jesuits and Galileo's Ideas of Science and Nature." *Annali dell'Istituto e Museo di Storia della Scienza di Firenze* (1983) anno 8, fascicolo 2:1–68.

Castelli, Benedetto. *Carteggio.* Massimo Bucciantini, ed. Florence: Olschki, 1988.

Cavalieri, Bonaventura. *Lo specchio ustorio. . . .* Bologna: 1632.

———. *Geometria indivisibilibus continuorum nova quadam ratione promota.* Bologna, 1635.

———. *Exercitationes geometricae sex.* Bologna, 1647.

———. *Geometria degli indivisibili.* Translated into Italian by Lucio Lombardo-Radice. Turin: UTET, 1966.

———. *Carteggio.* Giovanna Baroncelli, ed. Florence: Olschki, 1987.

Caverni, Raffaello. *Storia del metodo sperimentale in Italia.* Florence, 1891–1900. Reprinted, Bologna: Forni, 1970.

Celebrazione dell'Accademia del Cimento nel tricentenario della fondazione. Pisa: Domus Galilaeana, 1958.

Cohen, I. Bernard. *Revolution in Science.* Cambridge, Mass.: Belknap Press, 1985.

———. *The Birth of a New Physics.* New York: Penguin Books, 1985.

Cooper, Lane. *Aristotle, Galileo, and the Tower of Pisa.* Ithaca: Cornell University Press, 1935.

Crombie, Alistair C. *Robert Grosseteste and the Origins of Experimental Science 1100–1700.* Oxford: Clarendon Press, 1953, 1962.

Dati, Carlo. *Lettera a Filaleti di Timauro antiate della vera storia della cicloide, e della famosissima esperienza dell'argento vivo.* Florence, 1663.

Diaz, Furio. *Il granducato di Toscana: I Medici.* Turin: UTET, 1976.

Drake, Stillman. *Galileo at Work: His Scientific Biography.* Chicago: University of Chicago Press, 1978.

Drake, Stillman, ed. and trans. *Discoveries and Opinions of Galileo.* Garden City, New York: Doubleday, 1957.

Drake, Stillman, and Drabkin, I. E., eds. and trans. *Mechanics in Sixteenth-Century Italy: Selections from Tartaglia, Benedetti, Guido Ubaldo, & Galileo.* Madison: University of Wisconsin Press, 1969.

Duhem, Pierre. *Études sur Léonard de Vinci.* Published in three series. Paris: Hermann, 1906–1913.

———. *Le système du monde: Histoire des doctrines cosmologiques de Platon à Copernic.* 10 vols. Paris: Hermann, 1913–1959.

Fabroni, Angelo. *Lettere inedite di uomini illustri. . . .* 2 vols. Florence, 1773–1775.

Favaro, Antonio. *Galileo Galilei e lo studio di Padova.* 2 vols. Florence, 1883.

———. "Documenti inediti per la storia dei manoscritti Galileiani nella Biblioteca Nazionale di Firenze." *Bullettino di Bibliografia e di Storia delle Scienze Matematiche e Fisiche* (1885) 18:1–112, 151–230.

———. "Sul giorno della nascita di Galileo." *Memorie del R. Istituto Veneto di scienze lettere ed arti* (1887) 22:703–711.

———. "Sulla veridicità del 'Racconto istorico della vita di Galileo' dettato da Vincenzio Viviani." *Archivio Storico Italiano* dispensa 2, 1915. Florence: Tipografia Galileiana, 1916.

———. "Di alcune inesattezze nel 'Racconto istorico della vita di Galileo' dettato da Vincenzio Viviani." *Archivio Storico Italiano* dispense 3 e 4, 1916. Florence: Tipografia Galileiana, 1917.

———. *Amici e corrispondenti di Galileo.* 3 vols. Reprinted, Florence: Salimbeni, 1983.

Favaro, Antonio, and Carli, Alarico. *Bibliografia Galileiana (1568–1895).* Rome, 1896.

Feyerabend, Paul K. *Against Method.* New York: Schocken Books, 1978.

Finocchiaro, Maurice A. *Galileo and the Art of Reasoning: Rhetorical Foundations of Logic and Scientific Method.* Dordrecht: Reidel, 1980.

Finocchiaro, Maurice A., ed. and trans. *The Galileo Affair: A Documentary History.* Berkeley: University of California Press, 1989.

Galilei, Galileo. *Opere di Galileo Galilei.* . . . 2 vols. Carlo Manolessi, ed. Bologna, 1655–1656.

———. *Le opere di Galileo Galilei. Edizione Nazionale.* Antonio Favaro, ed. 20 vols. Florence Barbèra, 1890–1909. Reprinted 1929–1939, 1964–1966, 1968.

———. *Sidereus nuncius.* . . . Venice, 1610.

———. *Il saggiatore.* Rome, 1623.

———. *Dialogo . . . sopra i due massimi sistemi del mondo, tolemaico e copernicano.* Florence, 1632.

———. *Discorsi e dimostrazioni matematiche intorno a due nuove scienze attenenti alla mecanica & i movimenti locali.* Leyden, 1638.

———. *Dialogues concerning Two New Sciences.* Henry Crew and Alfonso de Salvio, trans. New York: MacMillan, 1914.

———. *Dialogue Concerning the Two Chief World Systems—Ptolemaic & Copernican.* Stillman Drake, trans. Berkeley: University of California Press, 1953, 1967.

———. *Two New Sciences.* Stillman Drake, trans. Madison: University of Wisconsin Press, 1974.

———. *On Motion and On Mechanics.* I. E. Drabkin and Stillman Drake, trans. Madison: University of Wisconsin Press, 1960.

Galluzzi, Paolo. "Evangelista Torricelli. Concezione della matematica e segreto degli occhiali." *Annali dell'Istituto e Museo di Storia della Scienza di Firenze* (1976) anno 1, fascicolo 1:71–95.

———. "L'Accademia del Cimento: 'gusti' del principe, filosofia e ideologia dell'esperimento." *Quaderni Storici* N. 48 (Dec. 1981) anno 26, fascicolo 3:788–844.

Galluzzi, Paolo, and Torrini, Maurizio, eds. *Le opere dei discepoli di Galileo Galilei. Edizione Nazionale: Carteggio.* 2 vols. Florence: Giunti-Barbèra, 1975–1984.

Garfagnini, Giancarlo, ed. *Firenze e la Toscana dei Medici nell'Europa del '500.* 3 vols. Florence: Olschki, 1983.

Garin, Eugenio. *Scienza e vita civile nel Rinascimento italiano.* Bari: Laterza, 1965.

Geymonat, Ludovico. *Galileo Galilei.* Novara: Einaudi, 1957.

———. *Galileo Galilei: A Biography and Inquiry into His Philosophy of Science.* Stillman Drake, trans. New York: McGraw-Hill, 1965.

Giacomelli, Raffaele. *Galileo Galilei giovane e il suo "De motu."* Pisa: Domus Galilaeana, 1949.

Giusti, Enrico. *Bonaventura Cavalieri and the Theory of Indivisibles.* Rome: Cremonese, 1980.

Guldin, Paul. *De centro gravitatis.* . . . 2 vols. Vienna, 1635–1640.

Hale, J. R. *Florence and the Medici: The Pattern of Control.* 1977, 1983. Reprinted, London: Thames and Hudson, 1986.

Hall, A. Rupert. *Ballistics in the Seventeenth Century: A Study in the Relations of Science and War with Reference Principally to England.* Cambridge: Cambridge University Press, 1952.

———. *From Galileo to Newton: 1630–1720.* London: Collins, 1963.

Kepler, Johannes. *Nova sterometria doliorum vinariorum, . . .* Linz, 1615.

Koestler, Arthur. *The Sleepwalkers: A History of Man's Changing Vision of the Universe.* London: Hutchinson, 1959.

Koyré, Alexandre. *Études galiléennes.* Paris; Hermann, 1939, Reprinted, 1966.

———. "La mécanique céleste de J. A. Borelli." *Revue d'histoire des sciences* (1952) 5:101–138.

———. *La révolution astronomique: Copernic, Kepler, Borelli.* Paris: Hermann, 1961.

———. *Metaphysics and Measurement: Essays in Scientific Revolution.* London: Chapman & Hall, 1968.

———. *The Astronomical Revolution: Copernicus—Kepler—Borelli.* R.E.W. Maddison, trans. Paris: Hermann, London: Methuen, Ithaca, New York: Cornell University Press, 1973.

———. *Galileo Studies.* John Mepham, trans. Hassocks, Sussex: The Harvester Press, 1978.

Kris, Ernst, and Kurz, Otto. *Die Legende vom Künstler: Ein historischer Versuch.* Vienna: Krystall Verlag, 1934.

———. *Legend, Myth, and Magic in the Image of the Artist.* Alastair Laing, trans. New Haven: Yale University Press, 1979.

Kuhn, Thomas S. *The Structure of Scientific Revolutions.* 2d ed. Chicago: University of Chicago Press, 1970.

———. *The Essential Tension: Selected Studies in Scientific Tradition and Change.* Chicago: University of Chicago Press, 1977.

Manno, Antonio, ed. *Cultura, scienze e tecniche nella Venezia del Cinquecento. Atti del Convegno Internazionale di Studio: Giovan Battista Benedetti e il suo tempo.* Venice: Istituto Veneto di Scienze, Lettere ed Arti, 1987.

McMullin, Ernan, ed. *Galileo: Man of Science.* New York: Basic Books, 1967. Reprinted, Princeton Junction: The Scholar's Bookshelf, 1988.

Meinel, Christoph. "Early Seventeenth-Century Atomism: Theory, Epistemology, and the Insufficiency of Experiment." *Isis* (1988) 79:68–103.

Mersenne, Marin. *Harmonie Universelle.* 3 vols. Paris, 1636. Reprinted, Paris: Centre National de la Recherche Scientifique, 1963.

———. *Les Nouvelles pensées de Galilée.* Paris, 1639. Critical edition with introduction and notes by Pierre Costabel and Michel-Pierre Lerner. 2 vols. Paris: Vrin, 1973.

Middleton, W. E. Knowles. *The History of the Barometer.* Baltimore: Johns Hopkins University Press, 1964.

———. *The Experimenters: A Study of the Accademia del Cimento.* Baltimore: Johns Hopkins Press, 1971.

Morpurgo-Tagliabue, Guido. *I processi di Galileo e l'epistemologia.* Milan: Edizioni di Comunità, 1963; Rome: Armando, 1981.

Moscovici, Serge. *L'expérience du mouvement: Jean-Baptiste Baliani, disciple et critique de Galilée.* Paris: Hermann, 1967.

Nardi, Antonio. "Censura sopra varii pensieri del Galilei." *La Critica Politica* (April–June 1990) 14-2:41–45.

Nelli, Giovanni Batista Clemente de'. *Vita e commercio letterario di Galileo Galilei.* Lausanne, 1793.

Olschki, Leonardo. *Geschichte der neusprachlichen wissenschaftlichen Literatur.* vol. 3: *Galilei und seine Zeit.* Halle (Saale): Max Niemeyer Verlag, 1927. Reprinted, Vaduz: Kraus Reprint, 1965.

Ornstein, Martha. *The Role of Scientific Societies in the Seventeenth Century.* First edition, privately printed, 1913. Chicago: University of Chicago Press, 1928, 1938. Reprinted, Hamden, London: Archon Books, 1963.

Procissi, Angelo. *La collezione galileiana della Biblioteca Nazionale di Firenze.* 2 vols. Rome: Istituto Poligrafico dello Stato, 1959–1985.

Redondi, Pietro. *Galileo Eretico.* Turin: Einaudi, 1983.

———. *Galileo: Heretic.* Raymond Rosenthal, trans. Princeton: Princeton University Press, 1987.

Rose, Paul Lawrence. *The Italian Renaissance of Mathematics: Studies on Humanists and Mathematicians from Petrarch to Galileo.* Geneva: Droz, 1975.

Santillana, Giorgio de. *The Crime of Galileo.* Chicago: University of Chicago Press, 1955. London: Mercury Books, 1961.

Schmitt, Charles B. *Studies in Renaissance Philosophy and Science.* London: Variorum Reprints, 1981.

Segre, Michael. "The Role of Experiment in Galileo's Physics." *Archive for History of Exact Sciences* (1980) 23:227–252.

———. "Torricelli's Correspondence on Ballistics." *Annals of Science* (1983) 40:489–499.

———. "Galileo as a Politician." *Sudhoffs Archiv* (1988) 72:69–82.

———. "Viviani's Life of Galileo." *Isis* (1989) 80:207–231.

———. "Galileo, Viviani and the Tower of Pisa." *Studies in History and Philosophy of Science* (1989) 20:435–451.

———. "Redondi's Theory and New Perspectives in Galilean Studies." *Archives Internationales d'Histoire des Sciences* (1990) 124:3–10.

———. "Science at the Tuscan Court, 1642–1667." In Sabetai Unguru, ed., *Physics, Cosmology and Astronomy, 1300–1700.* Dordrecht: Kluwer, 1991, pp. 291–304.

Settle, Thomas B. "An Experiment in the History of Science." *Science* (1961) 113:19–23.

Shapere, Dudley. *Galileo: A Philosophical Study.* Chicago: University of Chicago Press, 1974.

Shapin, Steven, and Schaffer, Simon. *Leviathan and the Air-Pump: Hobbes, Boyle, and Experimental Life.* Princeton: Princeton University Press, 1985.

Shea, William. *Galileo's Intellectual Revolution: Middle Period, 1610–1632.* New York: Science History Publications, 1972.

Targioni Tozzetti, Giovanni. *Notizie degli aggrandimenti delle scienze fisiche accaduti in Toscana nel corso di anni LX del secolo XVII.* Florence: 1780. Reprinted, Bologna: Forni, 1967.

Torricelli, Evangelista. *Opera geometrica.* Florence, 1644.

———. *Lezioni accademiche.* Florence, 1715.

———. *Opere.* Gino Loria and Giuseppe Vassura, eds., vols. 1–3. Faenza: Montanari, 1919; vol. 4, Faenza: Lega, 1944.

———. *Opere scelte.* Lanfranco Belloni, ed. Turin: UTET, 1975.

Torricelliana. Faenza: Unione Tipografica, 1945.

Torrini, Maurizio. *Dopo Galileo: Una polemica scientifica (1684–1711).* Florence: Olschki, 1979.

Vasari, Giorgio. *Le vite de' più eccellenti architetti, pittori e scultori italiani, da Cimabue insino a' tempi nostri.* Florence, 1550.

———. *Le vite de' più eccellenti pittori scultori, et architettori. . . .* Florence, 1568.

———. *Lives of the Artists.* A selection translated by George Bull. 2 vols. London: Penguin Books, 1965, 1971.

———. *Le vite de' più eccellenti pittori, scultori e architettori nelle redazioni del 1550 e 1568.* Rosanna Bettarini, ed., annotated by Paola Barocchi. 5 vols. of text, 2 vols. of notes. Florence: Sansoni, 1966.

———. *La vita di Michelangelo.* Paola Barocchi, ed. Milan, Naples: Riccardo Ricciardi, 1976.

Waard C. de. *L'expérience Barométrique: Ses antécédants et ses explication.* Thouard: Imprimerie Nouvelle, 1936.

Wallace, William A., ed. *Reinterpreting Galileo.* Washington, D.C.: Catholic University of America Press, 1986.

Wisan, Winifred L. "The New Science of Motion: A Study of Galileo's De motu locali." *Archive for History of Exact Sciences* (1974) 13 : 103–306.

Wohlwill, Emil. *Galilei und sein Kampf für die copernicanische Lehre.* 2 vols. Leipzig (1909–1926). Reprinted, Wiesbaden: Martin Sändig, 1969.

Yates, Frances A. *Giordano Bruno and the Hermetic Tradition.* London: Routledge and Kegan Paul, 1964.

Index